THE BODY'S BATTLE AGAINST POLLUTION

THE BODY'S BATTLE AGAINST POLLUTION

Dr. Radu Olinescu, Dr. Terrance L. Smith and Dr. J Hertoghe

NOVA SCIENCE PUBLISHERS, INC.
Commack, NY

Assistant Vice President/Art Director: Maria Ester Hawrys

Office Manager: Annette Hellinger

Graphics: Frank Grucci

Acquisitions Editor: Tatiana Shohov

Book Production: Ludmila Kwartiroff, Christine Mathosian, Joanne Metal and Tammy Sauter

Editorial Production: Susan Boriotti

Circulation: Cathy DeGregory, and Maryanne Schmidt

Library of Congress Cataloging-in-Publication Data

Olinescu, Radu

The body's battle against pollution / Radu Olinescu, Terrance L. Smith, J. Hertoghe.

      p.   cm.

Includes bibliographical references and index.

ISBN 1-56072-503-6

    1.   Environment toxicology. 2. Pollution—Health aspects. 3. Metabolic detoxication.

    4. Free radicals (Chemistry)-Pathophysiology. 5. Environmentally induced diseases.

    I. Smith, Terrance L., Dr. II. Hertoghe, J. III. Title

RA1226.045   1997   97-38186

615.9'02—dc21   CIP

CONTENTS

PART IV: OTHER POLLUTANTS AND FACTORS AFFECTING EXPOSURE TO POLLUTANTS

FORWARD

For much of the time since the industrial revolution, environmental pollution has been accepted as a necessary, if unfortunate, consequence of civilization. Tall smokestacks belching black smoke were seen as markers of progress, producing desired goods and wealth. Gradually people began to realize a high price was being paid for this form of progress. Toxic chemicals settled out of the air. The water was not only unfit to drink, but often smelled so bad that no one wanted to be near it. Oil spills coated beaches, killing fish, birds, and other sea creatures. The air, water and even the land were being destroyed and public opinion began to change. The smokestacks came to be viewed as dangers, threatening to destroy the environment and endangering the health of people. This perception was given even greater credibility by industrial disasters at places like Sevesso, Bhopal, Chernobyl, and the Minamata Gulf villages. Each of these accidental releases of toxic materials resulted in hundreds of people being killed and poisoned. With the breakup of the Soviet Union in 1989, the extent of pollution in Eastern Europe and Russia, which had been mostly hidden to Western eyes, became shockingly apparent. In Romania, for example, chemical pollution is widespread. Towns such as Giurgiu, Suceava, Baia Mare, Copsa Mica are so badly polluted that biochemical changes can be detected in the people there. This situation is typical of Eastern Europe.

There is no place on the earth free from the effects of chemical pollution and, in spite of great improvements in some parts of the world in controlling pollution, the burden of pollution continues to increase. Carbon emissions from the combustion of fossil fuels worldwide amount to 6 billion gigatons (6×10^{18} tons). Unless there is a change, this will increase to more than 25 billion gigatons (25×10^{18} tons) by the middle of the next century. The greenhouse effect of this carbon load has long been debated, but is now widely accepted. Another pollutant is chlorine, released to the atmosphere in the form of chloroflurocarbons from leaking air conditioners and refrigerators, and, at one time, from aerosol spray cans. While use of chloroflurocarbons is being curtailed worldwide, one molecule of chlorine can remain in the stratosphere from 75 - 110 years and break down 100,000,000 molecules of ozone. It has been reported that an estimated 100 tons of cadmium and similar amounts of arsenic and mercury have been released to the environment. The toxic effects of these elements include childhood leukemia, a disease on the rise in some parts of the world. There are some 65,000 known synthetic chemicals, but only 20% of them have been tested for their effects on people and animals.

While some parts of the world have made great improvement in their local environmental quality, other parts of the world have not even begun to address the problem. It can take a long time for public opinion to become strong enough to force change. Even then, there are many barriers to change: economic interests, political interests, including the resistance of totalitarian regimes, and lack of education on the part of the population. All too often change is not begun until there is a disaster, such as the industrial accidents mentioned earlier, attracting the attention of the news media, mobilizing worldwide opinion. By the time pollution reaches this level, where it is impossible to ignore, the removal of contaminated soil and overall cleanup has become difficult and expensive even for wealthy countries. Consider that the United States allocated 3 billion dollars in 1991 for cleanup of toxic waste dumps associated with weapons plants, military bases, and abandoned industrial sites.

Many books have been written, both scientific and popular, on the sources of pollution and on pollution's biological, economic, and social effects. This book addresses the question; if pollution is so widespread, how is it that life has been able to survive on earth? The answer is in the defensive systems of the body, which are able to give at least partial protection from the free radicals that result when the body metabolizes many types of chemical pollutants, and that even allow us some degree of adaptation. But, amazing as the body's defenses are, they can be overwhelmed. When this happens, a wide variety of diseases can result.

The authors have many years of experience as clinical and biochemical researchers into the defensive mechanisms of the body. This book is based on both the authors published research in the area of the body's defense against free radicals and on a careful search of the recent scientific literature.

PART I: POLLUTION AND FREE RADICALS

CHAPTER 1

POLLUTION AND ADAPTATION

1.1 NATURAL POLLUTION AND MAN-MADE POLLUTION

For thousands of years, mankind has tried to control nature. This has lead to many spectacular successes, but at a great price. Too often mankind's accomplishments have damaged the environment and exhausted resources through lack of understanding of consequences, or because consequences were ignored in the name of greed or politics. When the population of the earth was small and the technological base primitive, compared to today, pollution caused by civilization was minor and confined to a small area. As population and technology have increased, the potential severity and extent of pollution have likewise increased. Today, pollution is spread worldwide, and is leading to changes in living conditions, health, and even the climate.

While the word "pollution" generally calls to mind the effluvium of man and his technology, pollutants also arise from natural events. We can, therefore, distinguish between natural and man-made pollution.

NATURAL pollution has been present since the earth began and includes the main factors that have shaped the land and climate. Volcanic eruptions released gases and the lava altered the shape of the earth. Likewise, earthquakes, floods and erosion changed the shape of the land, the course of the rivers, and the mineral content of the seas. Even the release of oxygen to the atmosphere by early forms of life can be considered pollution. Of course, the existence of oxygen breathing plants and animals, including man, would not have been possible without this "pollution." Animal life present today continues to cause natural pollution through the release of methane from the digestive tracts of animals depending on fermentation for digestion, such as termites and cows. Life cannot exist without causing pollution. Living creatures remove resources, called nutrients, from the environment and release waste products back to the environment. However, in a normal ecological relationship, the wastes of one creature feeds another. Thus, in nature, there are a series of interrelated biological cycles. When these cycles are in balance, there is minimal pollution released to the environment.

MAN-MADE pollution, on the other hand, results from man's attempt to dominate and control nature, even if there is no intent to destroy or damage the environment. This form of pollution results from the activities of man, including agriculture, manufacture, transportation, and everyday living that produces domestic wastes. Like natural pollution, man-made pollution is a consequence of the existence of life. However, the intelligence and tool making ability of mankind greatly increases the potential impact.

Man's first activity that impacted the environment in a major way was probably the burning of forests to produce arable land. As early man developed agriculture, he became less dependent on the vagary of nature than he was as a hunter-gatherer. At the same time, man's impact on the environment was greatly increased. When the industrial revolution arrived, the impact on the environment was again greatly multiplied. It is estimated that in the past 50 years, the output of pollution generated by mankind has exceeded the pollution of the previous 500,000 years by a factor of 1,000.

During the past 20 years it has become apparent how severely this artificial pollution has impacted the environment. The evidence is seen in acid rain, the expansion of deserts, global warming, and the formation of the hole in the antarctic ozone layer.

It is indeed ironic that man's quest to improve his quality of life has lead to manufactured goods, chemicals, etc., which in turn, through the destruction of the environment and loss of natural resources, threaten to destroy his quality of life [50, 140].

1.2 THE MOVEMENT OF POLLUTION IN THE ENVIRONMENT

Modern man has come to accept that the earth has limited resources. What is often overlooked is that this means the elements that make up living things are recycled. Thus, oxygen, carbon, nitrogen, and the other elements found in the body pass through cycles, from living organisms, to the environment, back to living organisms, etc.

This unity of nature is summed up in the first of the ecological laws proposed by Dr. Barry Commoner, director for the Center for the Biology of Natural Systems, Queens College, New York, published in his book *Making Peace with the Planet* [47]. The first law states "everything is connected to everything else," describing the biological and ecological networks existing in nature. It emphasizes the interdependence of all life on earth.

As an example, consider the photosynthetic and respiratory cycles of plants and animals (Fig. 1). As described in figure 1.1A, plants use sunlight to drive photosynthesis, making sugars and oxygen from carbon dioxide and water. Plants and animals use the oxygen for respiration, giving up carbon dioxide and water in the process. As a result, oxygen, carbon, etc. are continually moving between plants, the atmosphere, and animals. Nitrogen is another important element for living things and it also goes through cycles (fig. 1.1B). Nitrogen, as part of rotting plant material and decaying animals, finds its way into the soil, where the plant material contributes to soil porosity [50]. Pollution interferes with these cycles directly by reacting with these elements, preventing their use, and indirectly by introducing toxic materials or by upsetting the balance between parts of the cycle. If the balance between parts of the cycle is pushed beyond its capability to recover, the induced change has repercussions throughout the ecological cycle. An example is the eutrophication of ponds, in which an excess of nitrogenous compounds causes excessive growth of aquatic plants and algae, in turn causing the pond to fill with sediment.

It only takes a little thought to come to understand that everything is indeed connected to everything else. It is also apparent that the ecological cycles and networks are very complex, involving many regulatory and feedback paths. The effect of altering the network at one point is difficult to forecast.

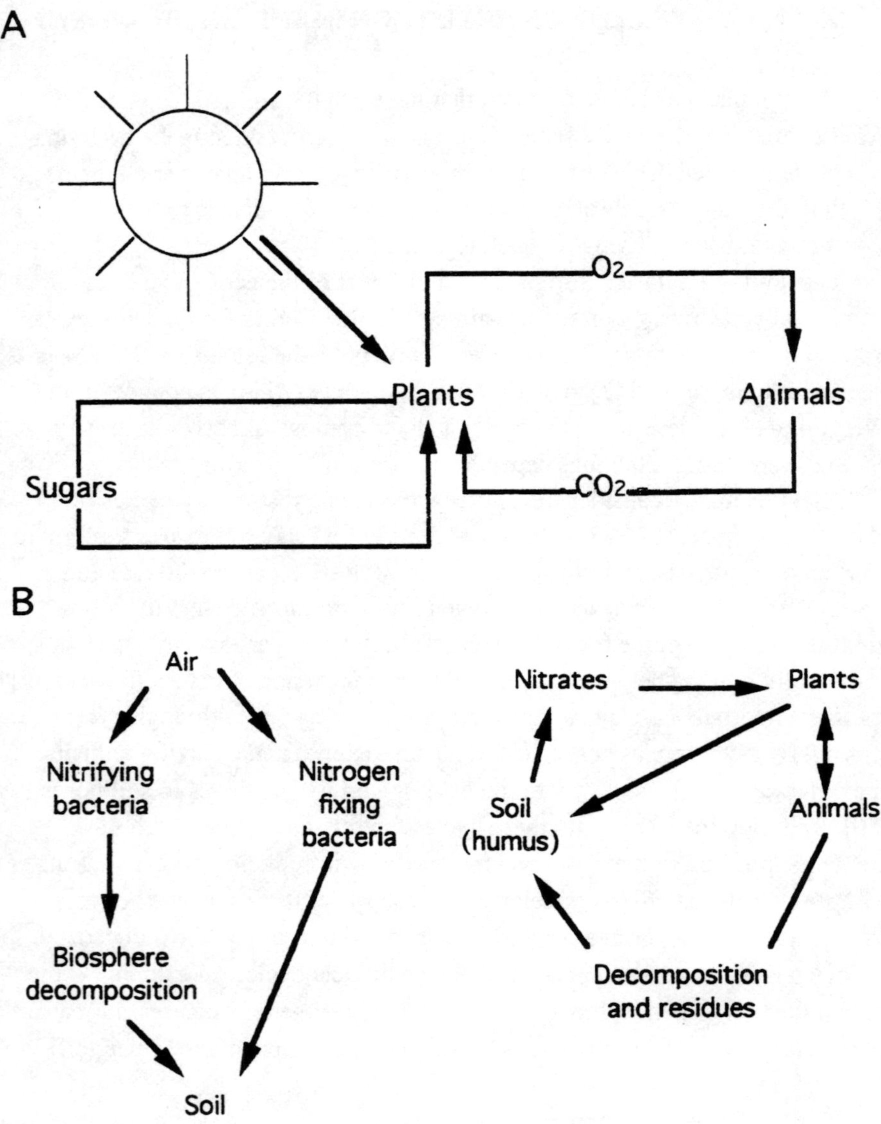

Figure 1.1.   A: The photosynthesis cycle and the role the solar energy in the formation of energy supplying compounds. B: The nitrogen cycle.

Most pollutants, especially metals like Pb, Hg, Cd, Cu, etc., and organic compounds like DDT also move through the environment. As figure 1.2A shows, gaseous wastes are washed out of the atmosphere by rain, contaminating the soil and water supplies. Contaminated soil and wastewater can be carried to the oceans. Even a locally polluting event has the potential to spread over a wide area. Pollutants in the air, soil, or water get into living things through the food chain (fig. 1.2B). Commoner's second ecological law addresses this worldwide spread of pollutants. It states "everything has to go somewhere." This is actually a restatement of the principle of conservation of mass, which might be stated as "everything can be transformed, but nothing is lost" and applied to pollution. All the by-products of man's activities are not lost, even when dispersed into the ocean. Sooner or later, pollutants, no matter how well dispersed, will enter the biological cycles. A typical example is DDT, a compound that is essentially nonmetabolizable, which can be detected in the ice covering Greenland and in marine organisms around the South Pole more than ten years after its use has been discontinued. Similar examples can be found for the world's great lakes and rivers.

The seriousness of pollution and the ease with which it spreads throughout the world is only now being appreciated. Serious attempts to control pollution on a worldwide scale have only begun in the past 10 years. However, the past 50 years have introduced new major sources of pollution, or increased the impact of old sources of pollution. Some of these global effects include:

- Release of radioactive material from atomic bomb tests and nuclear power plant accidents.
- A great increase in the atmospheric content of CO_2 from burning fossil fuels, contributing to the green house effect.
- Acid rain resulting from industrial gaseous wastes.
- Decrease in the ozone layer by released chlorinated fluorocarbons.
- Extinctions of animals and plants.

The effects of pollution are easily seen at the local level through:

- Nuclear accidents.
- Industrial and domestic waste released to rivers and lakes.
- Eutrophication of ponds and streams.

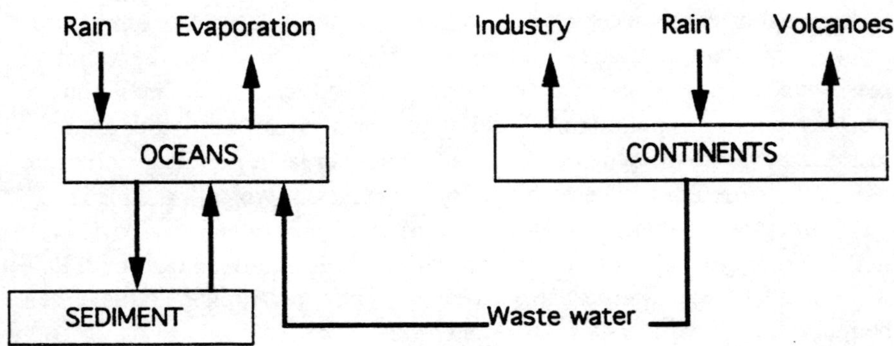

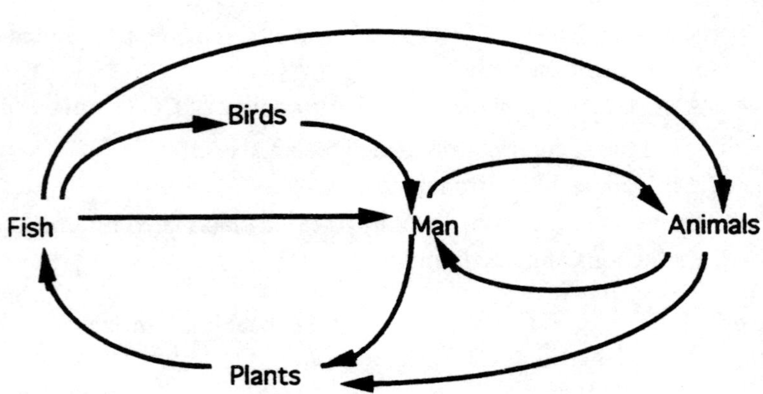

Figure 1.2. Cycles in nature. A: The circulation of metals (Hg, Pb) in the environment. B. The circulation of polluting compounds in the biosphere.

- Oil spills
- Bacterial and algal growth on pond surfaces and beaches.
- Mass toxic effects of industrial accidents (e.g., dioxin spills).

1.3 ADAPTATION TO POLLUTION

The history of life on earth has been one of continuous adaptation of living organisms to changing environmental conditions. According to Darwin's theory, that fight for survival was the driving force for the differentiation of the species. This process also aided adaptation by the creation of new systems for protection from toxic or otherwise harmful environmental factors. In this process of evolution and adaptation, those organisms that were able to develop protective mechanisms survived, the others died. This is reflected in Dr. Commoner's third ecological law; "nature knows best." During the evolution of life, nature had occasion to experiment with thousands of chemical combinations, but kept only a few as important to the composition and metabolism of organisms. These compounds can be synthesized through enzymatic reactions, and are degraded through other enzymatic actions.

Spores of bacteria and molds are able to survive extreme environmental conditions, including high temperatures, vacuum, and radiation, such as might be found in outer space. In 1988 it was found that some microorganisms are able to survive heating by producing "heat shock proteins." These stress proteins, which are produced enzymatically when induced by the stress, have also been found in microorganisms exposed to high levels of oxygen, to insecticides, and antibiotics and other chemicals.

Today, oxygen is essential for the dominant forms of life on earth. However, oxygen can also be considered the earth's first serious natural pollutant. Early forms of life are believed to have been unicellular organisms that obtained energy by breaking down chemicals found in the environment. Their metabolism required a reducing atmosphere, composed of ammonia, sulfurous gases, carbon dioxide, and traces of water. Eventually, these cells evolved the ability to photosynthesize, and oxygen began to be formed as a byproduct of their metabolism. Added to this was oxygen released from the degradation of silicates in the soil. As oxygen accumulated in the atmosphere, the existing organisms had to adapt by finding ways to protect themselves from the toxic effect of this molecule [211, 241]. So, new life forms evolved that used oxygen for respiration and but had biochemical mechanisms to protect them from oxygen toxicity [92]. An example of adaptation to toxic compounds that occurs today is the ability of microorganisms to resist antibiotics by the induction of enzymes that degrade them.

Adaptation, which is easy to study in simple organisms, occurs to some extent in all living things. Unfortunately, adaptation in animals, and plants is slow, while microorganisms exhibit a much more rapid ability to adapt. This is partly due to the greater metabolic rate of microorganisms, which is 100 to 1,000 times greater than that of fishes. Secondly, the reproductive cycle of the higher animals requires between 1 and 10 months, while for bacteria it is on the order of hours. Today, every year some 25,000 new compounds are synthesized in the laboratory. Some 500 of these eventually become widely used for commercial purposes, adding to the approximately 2,000,000 already known compounds. This large number of chemicals present in the environment and the slow rate of adaptation for higher forms of life cause many people, including scientists, to wonder if mankind will survive the increasingly dangerous environment presented by chemical pollution.

However, man does adapt to the environment and tolerates a certain level of contamination. For example, some adaptations to living at high altitude (modifications of glucose-6-phosphate dehydrogenase, an important enzyme in glucose metabolism, and polycythemia, an increase in the number of red blood cells in the circulation) are now part of the genetic code and are inheritable. People who are chronically exposed to chemicals, or who smoke or drink in moderation, induce greater levels of protective enzymes in their bodies, providing some protection from these toxic effects. As Barry Commoner has pointed out, the body of 20th century man includes Sr^{90} in the bones, I^{131} in the thyroid, DDT in the fats, and asbestos in the lungs, but he certainly lives better than his ancestors. As described in this book, the human body, like other organisms, is endowed with powerful resources of biochemical and immunological defense. These provide the ability to resist moderate levels of toxic materials. This may be the hope for man's survival as countries around the world initiate effective measures to limit or control pollution. If there is success in this endeavor, the amount of pollution in the world should be reduced to the point that man and other organisms are able to resist and adapt to the adverse effects of environmental pollution. Unfortunately, the success of this is not certain, since the maximal allowed level of pollutants varies widely between countries, and there are no international standards.

Most biochemical adaptations to environmental stress are not constantly turned on. Rather, when the proper conditions are present, the appropriate gene is activated, resulting in the synthesis of new or modified proteins. By increasing the synthesis of these proteins, cellular function is subtly altered, making the cells better able to cope with the environmental stress that triggered the change. Adaptive changes that occur at the genetic level are called genotypic adaptation. This form of adaptation makes use of the normally inherited variability in the genetic material that occurs through mutations and the action of natural selection.

The outward expression of genotypic adaptation that occurs throughout the organism's life span is called phenotypic adaptation. This is the visible or otherwise detectable change caused by the new or altered gene expression. Phenotypic adaptation, then, gives the organism the ability to resist the environment, especially when the environment is not ideally suited to that organism's life [63]. Phenotypic adaptation flows from environmental factors that activate specific genes, leading to increased biosynthesis of the coded molecule, altering some metabolic branches. Phenotypic adaptation is frequently reversible. An example is immunologic sensitization. When the environmental factor is no longer present, the modified gene will be turned off and the modified activity will decrease over a period of time.

Adaptive reactions involve a wide range of metabolic processes: ion transport across cell membranes, release of chemical energy by aerobic and anaerobic metabolic processes, etc. However, when the environmental factor is too great, exceeding the ability of the organism to adapt, symptoms of injury to the organism appear. These symptoms are characteristic of stress reactions, toxic effects, and decreased immunologic resistance. Some of these symptoms are characteristic of controversial diseases, such as chronic fatigue syndrome, chemical sensitivity, etc.

Adaptive processes may be classified by the time required to achieve a response. Short-term adaptation follows persistent, mild irritations. Short-

Table 1.1
Intensity of chemiluminesence (indicating oxidant stress) from lipids
extracted from the heart of rats exposed to different environmental
conditions (after Meerson [183])

Treatment	Chemiluminescence intensity (counts / sec / mg lipid)	Statistical significance
1. Control	16.0±1.5	
2. Stress	41.5±4.0	$p < 0.001$
3. Hyperoxia	21.0±1.6	$p < 0.1$
4. Hypooxia	24.2±2.7	$p < 0.05$

term adaptation makes use of existing physiologic mechanisms to respond to the environment. An example would be thermoregulation, in which the body increases heat production in response to cold.

Long-term adaptation involves a lasting change as might result from exposure to irritants, including pollutants. This adaptation involves the altered

gene expression discussed above. A comparison of Short-term adaptation and Long-term adaptation is shown in figure 1.4.

There is a constant interaction between gene expression and biochemical activity by feedback mechanisms. This feedback is under the control of hormones and secondary messengers, such as calcium ions and cyclic AMP. Enzymatic induction is a common mode of adaptation and has been observed in long-time drug users, alcoholics, smokers, and people chronically exposed to chemicals.

However, adaptation does not necessarily provide total protection from environmental factors. In one experiment, rats chronically exposed to high oxygen levels (hyperoxia), low oxygen levels (hypoxia), or stress had increased levels of lipid peroxide in their heart muscle (as measured by chemiluminescence) (Table 1.1). All of the treatments, including stress, produced damage as indicated by a significant increase in chemiluminescence. The animals used in this experiment had been adapted to the experimental conditions. Note that in this experiment adaptation reduced the damage to the heart muscle, but could not eliminate it. For nonadapted animals, the chemiluminescence values were twice those of the adapted animals [183, 213].

Since this condition, termed oxidative stress, is a common result of exposure to chemical pollution, it is apparent that an adaptive mechanism to such a challenge would be beneficial to survival. Adaptation to oxidative stress has been demonstrated in mammalian cells [294]. Oxidative stress can also result from intracellular redox reactions that occur during the body's attempt to detoxify certain pollutants. Severe oxidative stress can alter cellular function, compromising or destroying vital cell activities. Studies on bacteria have shown that these organisms can survive a very high level of oxidative stress if a relatively low-level stress is first applied. Bacteria that are first exposed to low concentrations of hydrogen peroxide (10 - 60 μM) or other superoxide generating agents survived longer than controls when then exposed to a normally lethal concentration. This resistance appears to be dependent on the expression of at least 30 genes inducing the synthesis of antioxidant enzymes, DNA repair enzymes, and other proteins. Catalase, glutathione peroxidase/reductase, and alkyl hydroperoxide reductase are all powerful antioxidant enzymes and are among the principle proteins produced by bacteria in adaptation to hydrogen peroxide exposure. The same increased antioxidant enzyme production is seen in human fibroblast cells exposed first to low concentrations of hydrogen peroxide (3 - 15 μM) followed by a lethal concentration (about 15 mM). The increased expression of antioxidant enzymes allows the fibroblasts to survive the normally lethal exposure to hydrogen peroxide [294]. It is believed that mammalian cells normally adapt to changing oxidative stress levels by transient overexpression of antioxidant enzymes and other protein genes involving over 20 products. A brief growth phase is also observed in successfully adapted cells. It is likely that cells adapt to these experimental conditions by conserving energy and protecting their DNA through supercoiling during the initial exposure to hydrogen peroxide, and then they redirect their transcription and translation machinery towards synthesis of antioxidant enzymes and proteins and repair enzymes.

These experiments [228, 294] illustrate what most of this book is about. The body possesses antioxidant protective factors and other enzymatic machinery designed to protect the body from the effects of oxidative stress.

The amounts of these enzymatic systems can be increased, providing limited adaptation to oxidative stress in organisms, including humans, who have been exposed to chemical pollutants. This is a major aspect of the body's battle against pollution.

CHAPTER 2

OXYGEN ACTIVATION, FREE RADICALS, AND THE BODY

2.1 OXYGEN ACTIVATION AND THE FORMATION OF FREE RADICALS

The discovery of oxygen activation in the sixties and seventies had the greatest impact on chemistry, biology, and medicine, but also affected many other branches of science including geophysics, agronomy, radiobiology, etc. Interest in oxygen activation has been so great that in the last 10 years 2 - 3 books per year, 3,000 - 4,000 scientific papers, and 2 new journals devoted to this topic have appeared. As we will discuss later, free radicals are frequently formed when the body is exposed to or metabolizes environmental chemicals. Since the major part of the book will involve discussing the defences against and consequences of exposure to free radicals, we will here review the chemistry of free radicals as they relate to biochemistry.

Oxygen is the most abundant element in the earth's crust and the second most abundant element in the biosphere. The concentration of molecular oxygen (O_2) in dry air is 21%. Even at this relatively low concentration, oxygen has been shown to be damaging to biological molecules and cell structure [34, 56, 63]. Our bodies use O_2 to oxidize carbon- and hydrogen-rich molecules to obtain the energy and heat necessary for life. During this process, O_2 is reduced to H_2O.

The stepwise reduction of oxygen (addition of electrons) leads to the formation of reactive oxygen species (ROS), some of which are free radicals. A free radical is defined as any chemical species having one unpaired electron and is represented with a raised dot ($^\bullet$). Oxygen is actually a biradical with two unpaired electrons of parallel spin in antibonding orbitals. Thus, oxygen has a unique structure and unusual properties.

If oxygen attempts to oxidize another atom or molecule by accepting a pair of electrons, both new electrons must be of parallel spin to fit the vacant orbitals. Most biological molecules are covalently bonded. In covalent bonds, the two electrons forming the bond have opposite spins. Because of this, the reaction of O_2 with biomolecules is considerably slowed, a point of considerable importance to aerobic life. To overcome this spin restriction, when conditions are suitable, oxygen will accept electrons one at a time, leading to the formation of ROS (reactions 1 - 4).

$$O_2 + e^- \xrightarrow{H} HO_2^{\bullet +} \xrightarrow{pH\,7.3} H^+ + O_2^{\bullet -} \qquad \text{(superoxide)} \qquad (1)$$

$$O_2^{\bullet -} + e^- \xrightarrow{2H} H_2O_2 \qquad \text{(hydrogen peroxide)} \qquad (2)$$

$$H_2O_2 + e^- \xrightarrow{Fe} OH^- + OH^\bullet \qquad \text{(hydroxyl)} \qquad (3)$$

$$OH^\bullet + e^- \xrightarrow{H} H_2O \qquad \text{(water)} \qquad (4)$$

Nature has used the peculiar properties of molecular oxygen with care, avoiding as much as possible the formation of ROS in high amounts. One method to accomplish this is to use other methods of oxygen reduction such as enzymes with a transition metal (Fe, Cu, Zn) at the active site. Another method is to use the absorption of a quantum of light to drive the formation of singlet oxygen ($^1O_2$). This electronically excited state of oxygen is used in metabolic or physiologic processes that occur in the chloroplast and the retina [34, 56].

The superoxide radical ($O_2^{\bullet -}$) is produced in numerous biological processes, in particular the electron transport chains of mitochondria and the endoplasmic reticulum. The production of superoxide by activated leukocytes and macrophages has also been extensively studied. Many studies indicate that $O_2^{\bullet-}$ is not particularly reactive in aqueous solutions. Its protonated form $(HO_2^{\bullet})$ is considerably more reactive both as a reductant and as an oxidant.

It is now well established that the superoxide radical produces a more damaging ROS through the "Fenton reaction" (7) or the "iron-catalyzed Haber-Wiss reaction" (5). The steps of these reactions are:

$$2O_2^{\bullet-} + 2H^+ \rightarrow H_2O_2 + O_2 \tag{5}$$

$$O_2^{\bullet-} + Fe^{3+} \leftrightarrow (Fe^{3+}O_2^{\bullet}) \leftrightarrow Fe^{2+} + O_2 \tag{6}$$

$$Fe^{2+} + H_2O_2 \rightarrow Fe^{3+} + OH^- + OH^{\bullet} \tag{7}$$

Equation (7) was discovered by Fenton in 1890. It shows that $O_2^{\bullet-}$ dismutates to form H_2O_2 and also reduces iron complexes or free iron to the ferrous state (6), making it available to the Fenton reaction. An intermediate perferryl complex is formed in this reaction. The Fenton reaction has been the subject of a 40-year debate over its existence *in vivo*.

The hydroxyl radical $(OH^{\bullet})$ is one of the major products of the radiolysis of water and is responsible for much of the damage done to tissues during radiation exposure. It is highly reactive and able to damage all biological molecules.

The significance of these reactions to this book is that during the biotransformation of xenobiotic compounds or chemical pollutants, oxygen activation takes place. Among subcellular fractions, the microsomes, where most of the biotransformation of xenobiotics occurs, are the richest source of ROS *in vitro* [49, 108]. In spite of a great deal of study, the mechanism of oxygen activation during biotransformation has not been completely explained. Neither catalase nor carotenes, which inhibit the formation of ROS, are able to inhibit the first phase reactions of biotransformation (see chapter 5 for a discussion of biotrasfromation chemistry). The formation of intermediate reactive compounds, such as arene oxides, epoxides, endoperoxides, NIH shift, etc., indicate the complexity and variety of biotransformations. *In vitro* experiments based on keeping iron in the reduced state (in the presence of NADPH and Fe^{3+} or ascorbate and Fe^{3+}) suggest the involvement of cytochrome P_{450} and dependent enzymes, especially in the presence of promoters for free radical formation.

Hydroxyl radical generated under physiologic conditions reacts with aromatic compounds at a diffusion-controlled rate, giving rise, mostly, to hydroxylated end products. Historically, aromatic hydroxylation as a "trap" for $OH^\bullet$ first occurred in humans using salicylate (2-hydroxybenzoate). The high doses of aspirin (acetylsalicylate) sometimes given to patients with rheumatoid arthritis result in concentrations of salicylate in synovial fluid that could conceivably trap some $OH^\bullet$. The attack by $OH^\bullet$ on salicylate produces two major hydroxylated products: 2,3-dihydroxybenzoate and 2,5-dihydroxybenzoate. The later product can be formed by the action of mammalian cytochrome P_{450} on salicylate, whereas 2,3-dihydroxybenzoate cannot. Consequently, formation of the later product may be an index of *in vivo* $OH^\bullet$ production [108]. Higher levels of 2,3-dihydroxybenzoate following aspirin administration have been observed in body fluids from patients with rheumatoid arthritis compared to normal controls or in rats treated with adriamycin then exposed to hyperoxia or irradiated. These observations are consistent with salicylate acting as a trap for $OH^\bullet$ *in vivo*.

Oxidative stress in cells and tissues involves increased generation of $O_2^{\bullet-}$ and H_2O_2. This can be achieved by: 1) raising the O_2 concentration higher than normal *in vivo*, 2) adding toxic compounds that increase intracellular oxidant formation (alloxan, paraquat, adramycin, etc.). or 3) activating a large population of phagocytes, which produce ROS to kill engulfed bacteria, but can cause tissue damage when they generate excessive amounts of ROS.

As described in equations (1) to (4), the activation of oxygen requires the presence of free or loosely complexed iron. Organisms take great care in the handling of iron so as to minimize the amount of free iron within cells and in extracellular fluids. However, oxidative stress can itself provide free iron for free radical reactions. Thus $O_2^{\bullet-}$ can mobilize iron from ferritin and H_2O_2 can degrade heme proteins to release iron. The bleomycin assay [109], introduced to measure iron ions available for free radical reactions in human body fluids, finds that physiologic levels of iron in the plasma are generally equal to zero and are rarely greater than 3 µmol/L except in patients with iron

Table 2.1
Evidence supporting the idea of increased oxidant stress in rheumatoid disease.

Observation	Reference
Increased lipid peroxidation in serum and synovial fluid.	Merry, et al. [184]
Increased exhalation of pentane resulting from peroxide decomposition.	Lunec & Alloran [172]
Depletion of ascorbate in serum and synovial fluid.	Yagi [300]
Increased formation of 2,3-DHB from salicylate.	Grootveld, et al. [100]
Degradation of hyaluronic acid.	Grootveld & Gutteridge [101]
Increased amount of uric acid (purine oxidation product).	Ames [5, 7]
Formation of fluorescent proteins.	Lunec & Alloran [172]

overload. By limiting the availability of catalytic metal ions, $O_2^{\bullet-}$ and H_2O_2 at physiologic concentrations will have limited adverse effects. This production can be bypassed by certain inflammatory diseases where the formation of ROS is increased.

The observations described in table 2.1 provide information concerning the formation of free radicals under pathological conditions such as occurs in arthritic diseases. The interpretation of data obtained by these and other methods is often difficult due to several factors. There is great variation in individuals regarding the activity of protective antioxidant systems and free radical forming systems. Figure 2.1 illustrates the complex pattern of damage that can occur. Increased free radical formation may occur because of excessive phagocytosis or the metabolism of chemical pollutants. Free radicals in human fibroblasts will cause damage to DNA. Free radicals trigger a sequence of biochemical events leading to GSH depletion, Ca^{2+} decompartmentalization, release of iron, lipid peroxidation, and, eventually, cell and tissue destruction.

2.2 PEROXIDATION

The formation of peroxides, especially lipid peroxides, is a consequence of oxygen activation, interconversion of ROS, and the exhaustion of natural protective systems. *In vivo*, as is true *in vitro*, the most favorable substrates

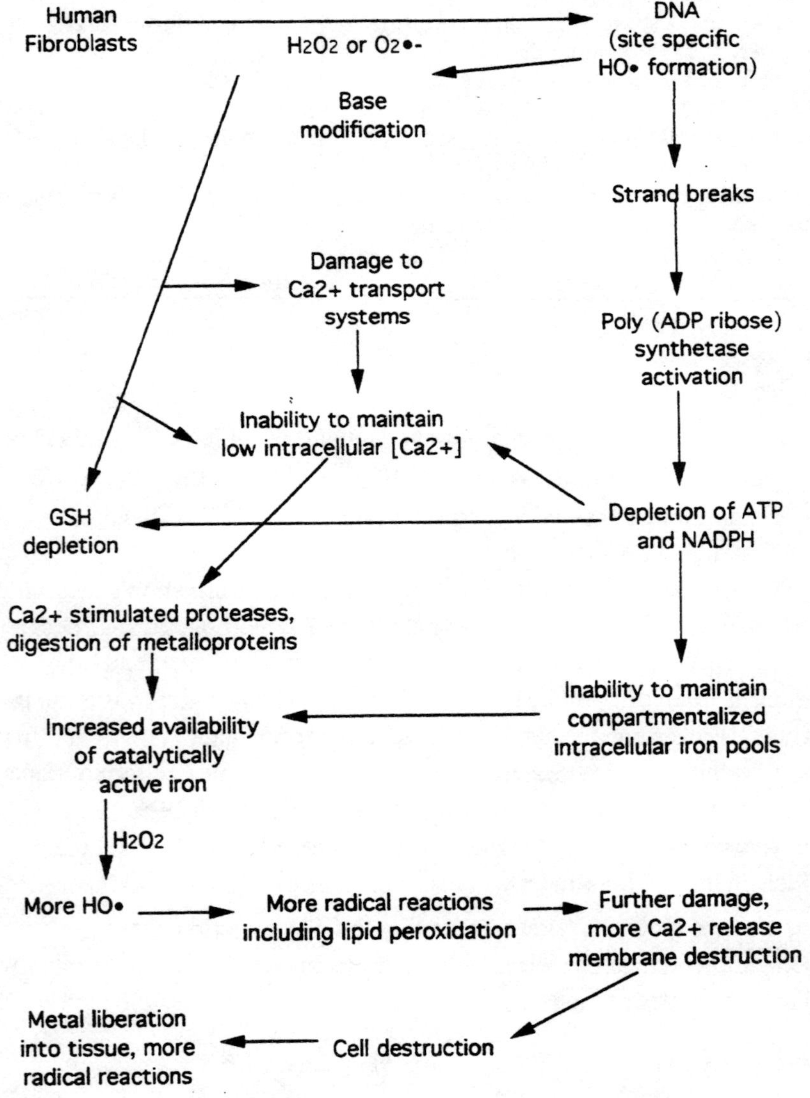

Figure 2.1.   Interrelationship of oxidant damaging mechanisms.

for peroxidation are the polyunsaturated fatty acids (PUFA), which are important components of cellular and subcellular membranes.

Lipid peroxidation occurs as a consequence of the formation of free radicals ($R^\bullet$) that have sufficient energy to abstract a hydrogen atom (H) from an unsaturated fatty acid (LH). The resulting carbon centered radical ($L^\bullet$) reacts rapidly with O_2 (10^9 - 10^{10} mol/sec) to form a peroxyl radical ($LO_2^\bullet$) and subsequently a lipid hydroperoxide (LOOH). Lipid peroxidation reactions have three phases: initiation, propagation, and termination.

$$LH + R^\bullet \rightarrow L^\bullet + RH \qquad \text{(initiation)} \qquad (8)$$

$$L^\bullet + O_2 \rightarrow LO_2^\bullet \qquad \text{(propagation)} \qquad (9)$$

$$LH + LO_2^\bullet \rightarrow LOOH + L^\bullet$$

$$L^\bullet + L^\bullet \rightarrow LL \qquad \text{(stable product)} \qquad \text{(termination)} \qquad (10)$$

$$LO_2^\bullet + LO_2^\bullet \rightarrow LOOL + O_2$$

Thus, the free radical chain reaction propagates until two free radicals destroy each other to terminate the chain of events (reaction 10). Hydroperoxides and cyclized endoperoxides are promoters for a new chain of lipid peroxidation by releasing $R^\bullet$ or free iron, which can enter the Fenton reaction.

$$LH + OH^\bullet \rightarrow L^\bullet + H_2O \qquad (11)$$

These reactions (8 - 11) only cover nonenzymatic lipid peroxidation. Enzymatic peroxidation refers to the generation of lipid peroxides at the active site of certain enzymes. The hydroperoxides and endoperoxides produced enzymatically are stereospecific and have important biological functions (oxidative turnover of biological membranes, biosynthesis of prostaglandins, etc.). The enzymes cyclooxygenase and lipoxygenase catalyze enzymatic peroxidation. Redox cycling of iron greatly increases the rate of lipid peroxidation. The activity of enzymes like cytochrome P_{450} reductase in the presence of NADPH can reduce iron complexes, as can the superoxide generated by the activity of xanthine oxidase [56, 108, 136].

A great variety of active biological compounds are produced by enzymatic oxidation of the PUFA, arachidonic acid. Some of the products (prostaglandins and leukotrienes) derive from PGG_2 (prostaglandin G_2), and endoperoxide, while others derive from HPETE (hydroperoxyeicosatetraenoic acid), an acyclic hydroperoxide. In prostaglandin biosynthesis the formation of free radicals and lipid peroxides is beneficial to the organism.

It is also possible for prostaglandin synthesis to be coupled with adverse free radical products. The enzyme prostaglandin H synthase is widely found in cellular membranes and catalyzes the formation of bicyclic hydroperoxides. This enzyme primarily acts as a dioxygenase but may also catalyze the reduction of PGG_2 into PGH_2, an endohydroperoxide. In the later reaction, prostaglandin H synthase is acting as a peroxidase.

During the peroxidase reaction other compounds may be involved:

$$\text{Arachidonic acid} \xrightarrow{\ cyclooxygenase\ } PGG_2 \tag{12}$$

$$PGG_2 + DH_2 \xrightarrow{\ PG\ H\ synthase\ } PGH_2 + D \tag{13}$$

where DH_2 is a reductant, or electron donor. The cooxidation of polycyclic aromatic hydrocarbons, aflatoxin B, and aromatic amines has been proven experimentally to occur during the biosynthesis of prostaglandins. The activation of polycyclic aromatic hydrocarbons by prostaglandin H synthase provides an additional way (additional to the pathway involving cytochrome P_{450} and NADPH, see chapter 5) for producing mutagenic and carcinogenic compounds. Any biochemical free radical generating system, such as Fe-ascorbate or microsomes Fe-NADPH, may initiate the epoxidation of polycyclic aromatic hydrocarbons and their subsequent binding to DNA [169]. The beneficial action of retinoids (13-cis-retinoic acid) in the treatment of cancer is probably due to the competitive inhibition of these reactions.

Nitrogen oxides (NO_2 and NO) are free radical promoters and are found in great amounts in polluted air. Tobacco smoke contains approximately 1000 ppm NO, which is gradually transformed into NO_2. It is proven that NO_2 attacks PUFA in the lungs resulting in increased lipid peroxidation. The reaction consists of several steps:

$$NO \longrightarrow O_2^{\bullet} \; NO_2 \xrightarrow{\;PUFA\;} L^{\bullet} \xrightarrow{\;O_2\;} LOO^{\bullet} \xrightarrow{\;NO\;} LO^{\bullet} \tag{14}$$

The reactive species NO_x may also react with H_2O_2 to produce the hydroxyl free radical:

$$NO + H_2O_2 \rightarrow OH^{\bullet} + HNO_2 \tag{15}$$

$$NO_2 + H_2O_2 \rightarrow OH^{\bullet} + HNO_3 \tag{16}$$

The relatively high concentrations of NO_x in tobacco smoke also stimulate pulmonary macrophages, which release H_2O_2, feeding reactions 15 and 16. Experimental evidence for the reactions described above was provided by the hydroxylation of phenols by a mixture of NO_2 and H_2O_2 (see reaction 14).

In spite of the dangerous peroxidative activity of NOx, organisms need these reactive species. Nitric oxide (NO) is also known as Endothelial-Derived Relaxing Factor (EDRF) and is produced in large quantities in the body following endothelial cell stimulation by acetylcholine. EDRF diffuses into the neighboring smooth muscle, producing relaxation of muscular tonus. EDRF mediates the vascular relaxation induced by some compounds such as adenine nucleotides, thrombin, bradykinin, and PUFA. This makes EDRF a second messenger for vasodilation. It acts synergistically with prostacyclin in inhibiting platelet aggregation. The action of EDRF is modulated by $O_2^{\bullet-}$ generating systems like xanthine oxidase.

Nitric oxide is produced in tissues by a specific synthase with the release of citrulline. Citrulline concentrations are significantly higher in rectal biopsies from patients with ulcerative colitis. As NO may cause tissue damage and increase the formation of carcinogenic nitrosamines, it is a possible link for the increased frequency of colonic carcinoma in cases of ulcerative colitis.

Bilirubin is another source of peroxidation. This compound is produced in the physiologic degradation of hemoglobin. The toxicity of bilirubin results in neurological damage to newborns if it is allowed to remain at a high level for too long a period [212]. Bilirubin is strongly lipophilic. It is known to associate with phospholipids and may catalyze the peroxidation of purified PUFA or tissue homogenates [274]. Peroxidation from this source is seen during liver failure. As shown in table 2.2, in cases of liver injury, the level of peroxidation increases in parallel with the level of bilirubin and glutamate-

Table 2.2
The level of lipid peroxides formed in some diseases. After Olinescu [212]

Disease	Lipid perox. as MDA[a] (μmol/L)	Bilirubin (mg %)	GP Transaminase (U/L)	GSH (mg %)
Control	3.6±3.8	1.05±0.8	26.5±3.5	47.4±4.3
Alcoholic hepatitis	16.8±5.4**	5.8±3.7*	38.5±4.1	32.5±1.6
Viral hepatitis	21.9±3.2**	9.3±4.2**	496.3±24.3**	26.9±2.8*
Hepatic coma	66.8±8.4**	19.9±2.6**	532.7±36.5**	24.6±3.9*
Hepatic cirrhosis	17.5±4.3**	15.4±5.2**	37.8±3.9	25.7±4.7*
Newborn icterus	12.3±2.6**	10.2±4.5**	—	25.7±4.7*
Arthritis	11.8±5.7*	1.15±0.9	28.8±4.8	32.7±5.5*

[a] Malondialdehyde
* $p<0.05$
** $p<0.01$

pyruvate transaminase. At the same time a significant reduction in reduced glutathione occurs. Measurement of reduced glutathione is often used as an index of the antioxidative systems in a tissue. The nonsignificant increase in lipid peroxidation commonly seen in moderate liver damage illustrate the resilience and adaptability of this organ.

The biotransformation of chemical pollutants occurs primarily in the liver. Injury (trauma, radiation, or chemical) to the liver produces tissue damage with cellular lysis and the release of PUFA and intracellular proteins. Hepatotoxicity caused severe chemical exposure can overcome the protective, antioxidant systems.

Heme oxygenase (HO) catalyzes the reaction:

$$\text{Heme} + \text{NADPH} + \text{O}_2 \xrightarrow{HO} \text{Biliverdin} + \text{NADP} + \text{CO} + \text{Fe}^{3+} \qquad (17)$$

Under physiological conditions, biliverdin is quickly converted to bilirubin by biliverdin reductase:

$$\text{Biliverdin} + \text{NADPH} \rightarrow \text{Bilirubin} + \text{NADP} \qquad (18)$$

Oxygen activation occurs by equation 19, catalyzed by cytochrome reductase:

$$\text{Heme-Fe}^{3+} + \text{NADPH} + O_2 \rightarrow \text{Heme-Fe}^{2+} + \text{NADP} + O_2^{\bullet-} \qquad (19)$$

Superoxide is converted to hydrogen peroxide spontaneously or by the action of superoxide dismutase (SOD):

$$2O_2^{\bullet-} \xrightarrow{\;SOD\;} H_2O_2 \xrightarrow{\;O2\ \&\ Fe\;} OH^{\bullet} + OH^- + O_2 \qquad (20)$$

Therefore, any condition in which there is large amount of hemoglobin degradation creates favorable conditions for peroxidation.

Metallic ions, such as Fe, Cd, Sn, Pb, and Mn, act by inducing Heme oxygenase activity, lipid peroxidation, and the oxidation of thiol compounds such as GSH. The rate of NADPH dependent peroxidation in liver microsomes at 37° C is approximately 240 nmoles O_2/mg protein and has a Km of 0.55 M [49].

As seen in table 2.3, lipid peroxidation is a nonspecific event and is a possible consequence of any toxic, infectious, traumatic, or diseased condition (diabetes, atherosclerosis, rheumatoid arthritis, etc.) [56, 108, 300]. The level of lipid peroxidation indicates the balance between oxidative stress and antioxidative, protective systems.

Recent studies suggest that for most toxic compounds, induction of lipid peroxidation is not the initial mechanism by which they cause cell injury. For example, paraquat causes cellular injury by stimulating the production of $O_2^{\bullet-}$ and H_2O_2. Lipid peroxidation does not occur until a later stage. As shown in figure 2.1, oxidant injury can lead to GSH depletion and the release of metal ions (Fe), leading in turn to lipid peroxidation. Peroxidation may even be delayed to the point of cell death and membrane lysis. Lipid peroxidation end products may be good markers of cellular injury in diseases like rheumatoid arthritis because it occurs as a consequence of the injury, rather than being the cause. Increased amounts of lipid peroxides and lipid peroxidation end products (aldehydes, fluorescent lipofuscins, etc.) can be detected in many diseased states since damaged cells and tissues peroxidize more rapidly regardless of the cause of injury, with the exception of tumors (table 2.4).

Table 2.3

Biochemical variations in rat liver from animals exposed to acute chemical intoxication or irradiated with a lethal dose. After Olinescu [212]

Exposure	Lipid perox. as MDA[a] (nmoles/mg protein)	Heme oxygenase (nmoles bilirubin/mg protein)	GSH (nmoles SH/mg protein)
Control	4.12±0.18	1.83±0.12	5.83±0.40
Cadmium (1.2 mg/kg)	5.21±0.16	3.24±0.14	4.38±0.15
Cadmium (5.0 mg/kg)	6.38±0.14*	5.87±0.18**	3.47±0.37*
Phenyl hydrazine (50 mg/kg)	6.59±0.21*	3.24±0.17	4.63±0.35
Phenyl hydrazine (100 mg/kg)	9.85±0.27**	6.58±0.26**	3.12±0.25*
Irradiation 10 Gy	7.65±0.32*	4.76±0.18*	3.16±0.27*

[a]Malondialdehyde
*p<0.05
**p<0.01

Table 2.4

Pathological conditions in which lipid peroxidation in blood has been found. After [25, 109, 211, 237, 300]

Condition	Extent of Peroxidation	Organ involved
Hyperbaric oxygen	+	Brain
Burns, transplants	+	Skin
Cataract	+ → +++	Eyes
Chemical toxicity	+ → +++	Liver
Metal toxicity	+ → +++	Liver
Anthracycline antibiotics	+ → ++	Heart
Asbestosis	+	Lungs
Nitryloacetate intoxication	+ → +++	Kidneys
Insecticides exposure	+ → +++	Liver
Endotoxin in liver injury	+ → +++	Liver
Styrene acrylamide	+ → +++	Liver
Hemodialysis	+ → +++	Erythrocytes
Post ischemia	+ → +++	Brain, heart
Muscular dystrophy	+ → ++	
Multiple sclerosis	+ → ++	

2.3 BIOLOGICAL CONSEQUENCES OF OXYGEN ACTIVATION

Biological transformation of chemical pollutants frequently involves the formation of free radicals or reactive oxygen species. The more reactive a free radical is, the smaller will be its range of diffusion. For example, OH$^\bullet$ and CCl$_3^\bullet$ have a half-life in a biological medium of 5 - 7 μsec, and their diffusion distance is about 80-nm [34]. Consequently, the type of damage a free radical produced in the endoplasmic reticulum may do to a cell depends on the radical's reactivity. If a free radical is to reach the nucleus with the possibility of binding covalently to DNA and causing cancer, the radical must be of only moderate reactivity. Highly reactive radicals will be more likely to initiate lipid peroxidation near the site they are formed. In this way, peroxides and their decomposition products (mostly aldehydes such as carbonyls, 4-hydroxylalchenals, etc.) may cause cell structural damage at relatively long distances from the site of the initial free radical formation.

2.3.1 MOLECULAR MODIFICATIONS; RADIATION INDUCED HYDROLYSIS

While it is clear biochemically that free radical formation causes alterations at the molecular level, it is often more difficult to trace the exact effects clinically. As seen in figure 2.2, oxygen activation proceeds from reactive oxygen species to peroxides and carbonyls. The damaging effects may occur immediately with the formation of a reactive oxygen species, or later when peroxides appear. If the damage does not occur until the formation of peroxides, there can be an amplification effect through the loss of the membrane barrier function towards calcium and other ions. If the calcium content of the cell is allowed to rise, calcium-dependent proteases (e.g., calpain) will be activated, which can stimulate other enzymatic reactions. Xanthine dehydrogenase is important in the degradation of purines. Oxidation of a cysteine residue or proteolytic cleavage of a certain peptide bond converts xanthine dehydrogenase into xanthine oxidase:

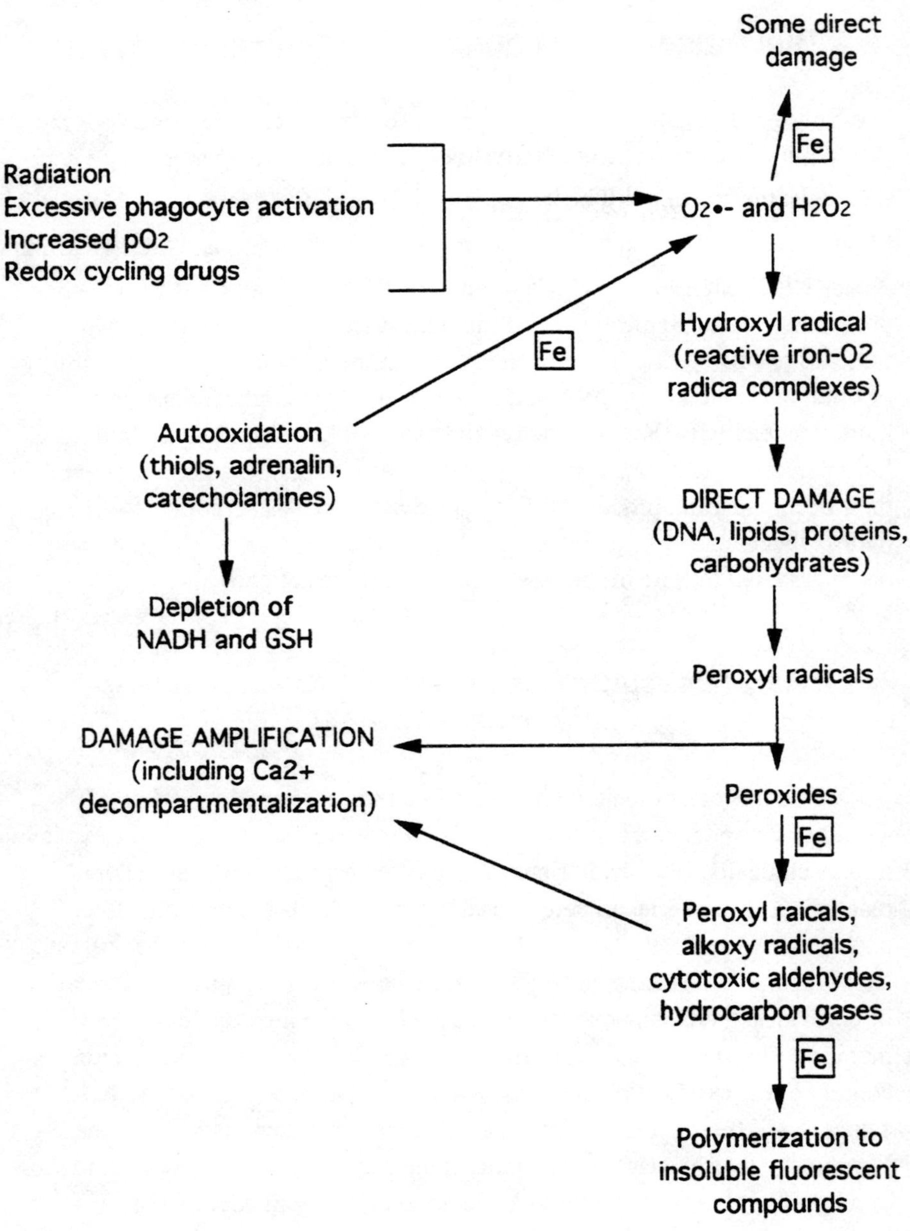

Figure 2.2.   Mechanism of cellular damage in oxidative stress.

$$\text{Xanthine dehydrogenase} \xrightarrow{\textit{calpain}} \text{Xanthine oxidase} \tag{21}$$

This conversion occurs during purification of the enzyme and during certain *in vivo* situations such as during a heart attack. Xanthine oxidase mediates the same reactions as xanthine dehydrogenase, but produces hydrogen peroxide as a byproduct:

$$\text{Xanthine} + O_2 \xrightarrow{\textit{xanthine oxidase}} \text{Uric acid} + H_2O_2 \tag{22}$$

Thus, reaction (22) can occur as a physiologic reaction in the degradation of purines and is a source of reactive oxygen species.

During exposure to ionizing radiation (α, β, and γ rays), reactive oxygen species are formed by the radiolysis of water. When this reaction occurs in tissues, there is ample opportunity for the free radicals and reactive oxygen species to attack proteins, lipid membranes, or to have an amplified effect through the reaction with Fe^{2+} (equations 5 - 7).

The oxygen effect in radiobiology is the radiation-induced damage that occurs in the presence of oxygen through its activation. Following the activation of oxygen, the first effect of exposure to ionizing radiation is often lipid peroxidation, both through direct peroxidative reactions and by the release of iron.

As shown in figure 2.3, oxidative stress produced from different sources give the same effects. These effects begin at the molecular level and quickly move to the cellular and tissue levels. The complexity of these events can be seen from figures 2.2 and 2.3. It should also be noticed that several pathways and mechanisms of oxygen activation may be involved.

Under acute or chronic radiation exposure, an increase in oxidation of various substrates (PUFA, purines, etc.) will occur, causing the elevation of peroxides in the system. Since radiation and chemical toxicity effects share several pathways, including oxygen activation, an increased sensitivity to chemical pollutants is to be expected in people with chronic radiation exposure. As has been shown, H_2O_2 is a compound common to the production of mutagenic, clastogenic, and other damaging endogenous compounds [6, 250].

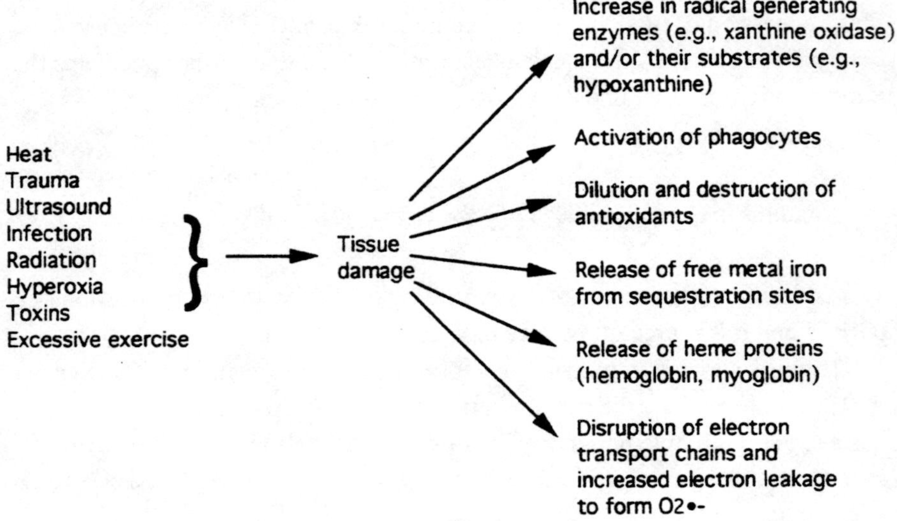

Figure 2.3. How tissue damage can cause oxidative stress.

2.3.2 CELLULAR MODIFICATIONS; PHAGOCYTOSIS

As mentioned, oxygen activation and the formation of reactive oxygen species occur predominantly in the biomembrane. As shown in figure 2.3, when peroxidation takes place, subsequent reactions proceed from molecular to cellular levels of damage and favor structural modifications, especially in the membrane:

a) Structural alterations to the lipid bilayer and the embedded proteins, including pores that penetrate the membrane;
b) Structural changes to PUFAs and phospholipids, making qualitative changes in the composition and function of the membrane. Such changes have been observed in chronic intoxications (by solvents, alcohol, etc.) and in some diseases (diabetes mellitus, cancer, atherosclerosis, etc.);

c) Effects 'a' and 'b' lead to a decreased membrane permeability and an alteration in the active transport of ions (Na, K, Ca, etc.);

d) These combined effects result in changes in ion concentrations inside the cell. Altered ion concentrations activate lytic enzymes. The release of cellular contents is responsible for the appearance of intracellular enzymes (transaminase, lactate dehydrogenase, etc.) in the blood during cellular and tissue injury.

Phagocytosis is a most illustrative example of oxygen activation and its effects on the cellular level, extending to the entire body. Phagocytosis is a complex and essential part of cell mediated immunity. Phagocytic cells in the blood include neutrophiles, monocytes, and eosinophils. Phagocytic cells in the tissues include macrophages and some epithelial cells. Briefly, the process involves attraction of a phagocytic cell to a foreign body through chemotaxis, recognition of the foreign body, adherence of the foreign body to the membrane of the phagocytic cell, engulfing the foreign body into a phagosome, and the destruction of the phagocytized organism through the combined action of proteases and reactive oxygen species.

a. This sequence of events is initiated by the presence of a gradient of a chemotactic substance. The chemotactic substance is generated through the interaction of the pathogen (bacteria, mineral dust, or altered endogenous structures) and the components of humoral immunity (plasma opsonins and complement). Chemotactic substances attract the attention of phagocytic cells, which follow the chemical trail up the concentration gradient to the target.

b. As phagocytic cells move up the concentration gradient, a sequence of biochemical and immunologic events occur within the cell and on its surface. Receptors are localized on the cell surface. The membrane becomes hyperpolarized, modifying calcium flux. Prostaglandin synthesis from arachidonic acid (obtained from PUFAs found in the membrane) also increases.

c. Oxygen activation is the most important step and occurs concurrently with the activation of the phagocyte. It involves a sequence of biochemical reactions that create and release reactive oxygen substances:

- The activity of glycoloysis increases to provide needed energy and the hexose monophosphate shunt increases to provide NADPH (as a coenzyme).

- Certain enzymatic reactions become active:

$$NADPH + 2O_2 \xrightarrow{\textit{oxidase}} NADP + 2O_2^{\bullet-} \qquad (23)$$

$$2O_2^{\bullet-} \xrightarrow{\textit{SOD}} H_2O_2 + O_2 \qquad (24)$$

$$H_2O_2 + Cl^- \xrightarrow{\textit{myeloperoxidase}} HOCl + OH^- \qquad (25)$$

- These enzymatic reactions assure a flux of reactive oxygen species that will kill most pathogens.
- The destructive action of reactive oxygen species and hypochlorite (HOCl) will be complemented by the activity of calcium dependent proteases and other bactericidal, anti-viral, and tumoricidal compounds.

d. The release of reactive oxygen species along with a number of other products from activated leukocytes (leukotrienes, cytokines, prostaglandins, etc.) produces a variety of consequences in the vicinity of the infected site. This inflammatory process is described in figure 2.4. As seen in this figure, this consequence of this process affects the entire body, depending on its magnitude.

A large number of natural and exogenous compounds interact with the process of phagocytosis. Anti-inflammatory compounds, whether natural (cortisone) or exogenous (bee venom, or drugs such as phenylbutazone, aspirin, etc.) inhibit phagocytosis and the longer ranged events. Other events that occur during phagocytosis include chemiluminescence, which can be measured *in vitro* using a luminometer. This physicochemical method has been used to test the action of anti-inflammatory drugs and other compounds [213].

The interaction of chemical pollutants with phagocytosis is also complex. At low levels, most xenobiotic compounds stimulate oxygen activation by a combined action on phagocytic membranes. But at higher concentrations, an opposite effect is seen, depending on the structure. The structure of chemical pollutants influences the threshold where these effects take place. Thus, while mineral dust, powders, fibers, exhaust gases, and tobacco smoke stimulate phagocytosis [56, 129, 213], uranium oxide dust and asbestos fibers are cytolytic towards pulmonary macrophages [256, 268]. A similar stimulatory effect on phagocytes is also exerted by aromatic compounds, such as phenols

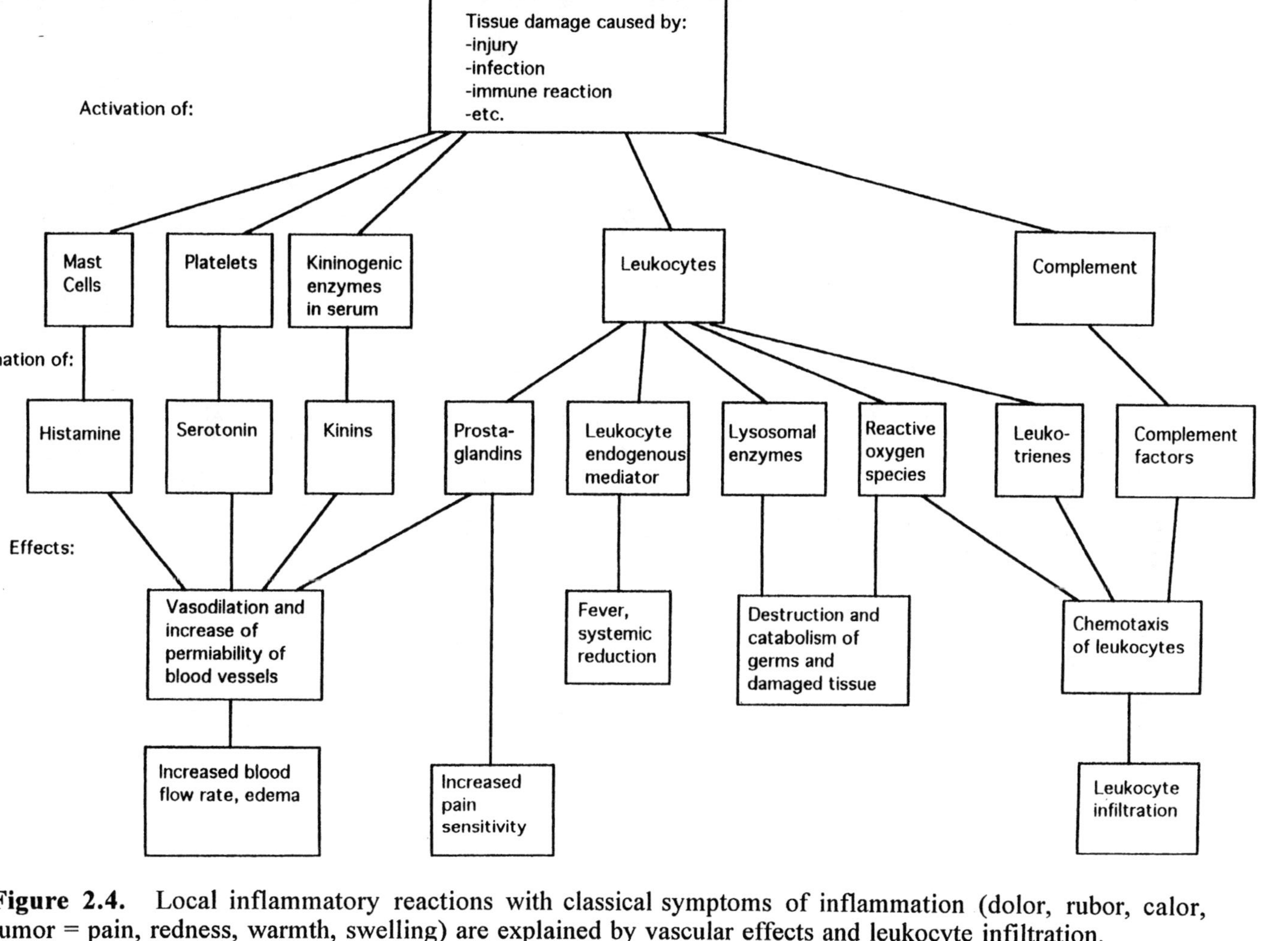

Figure 2.4. Local inflammatory reactions with classical symptoms of inflammation (dolor, rubor, calor, tumor = pain, redness, warmth, swelling) are explained by vascular effects and leukocyte infiltration.

and naphtols, but this effect may be lost because activated phagocytes may hydoxylate these compounds, at least *in vitro* [255, 278].

Scientists are very interested in the compounds released during phagocytosis. These compounds possess a great variety of structures and actions, and include histamine, prostaglandins, leukotrienes, proteases, thromboxans, etc. (see figure 2.4). Some of these compounds can cause delayed effects in the vicinity of an infection, such as the oxidative degeneration of collagen or hyaluronic acid. These effects may lead to rheumatism or collagenosis [100]. Interactions with collagen are complex. Macrophages release inhibitors of collagenase but at the same time, activate latent collagenase [221, 247]. Furthermore, collagen or its degradation products is able to stimulate phagocytes [213]. Thus, there can be unexpected consequences of phagocytosis, explaining the unpredictable clinical development of some diseases.

As seen in figure 2.5, injury following a bacterial invasion or exposure to a xenobiotic compound spreads from the initial site, unless effective protective or repair systems are in place. The fate of a cellular population depends on the balance between damaging and protective systems. Oxygen activation and its consequences form a significant part of the events described in figures 2.4 and 2.5. The cell damaging compounds are mainly reactive oxygen species and the protective systems consist mostly of enzymatic and nonenzymatic antioxidants. Gemas [91] has pointed out that the interaction between oxygen activation and the immune process is complex because most of the chemotactic factors, prostaglandins, leukotrienes, and cytokines are products of phagocytosis. The chemical messages exchanged among various immunocompetent cells during the inflammatory process also consist largely of products of oxygen activation [63, 119].

An apparently paradoxical action of phagocytic cells appears in carcinogenesis. Normally, leukocytes are able to destroy tumor cells [165], but the injection of an *in vitro* reactive oxygen species-generating system (such as xanthine + xanthine oxidase or stimulated leukocytes) into rats significantly increases the frequency of tumors [209]. In the same vein, stimulated leukocytes are able to activate polycyclic aromatic hydrocarbons [26]. Stimulated macrophages produce tumor necrosis factor (TNF) to activate phagocytic cells, as well as lipoprotein lipase, prostaglandins, and other compounds [31]. TNF is so toxic it cannot be used clinically.

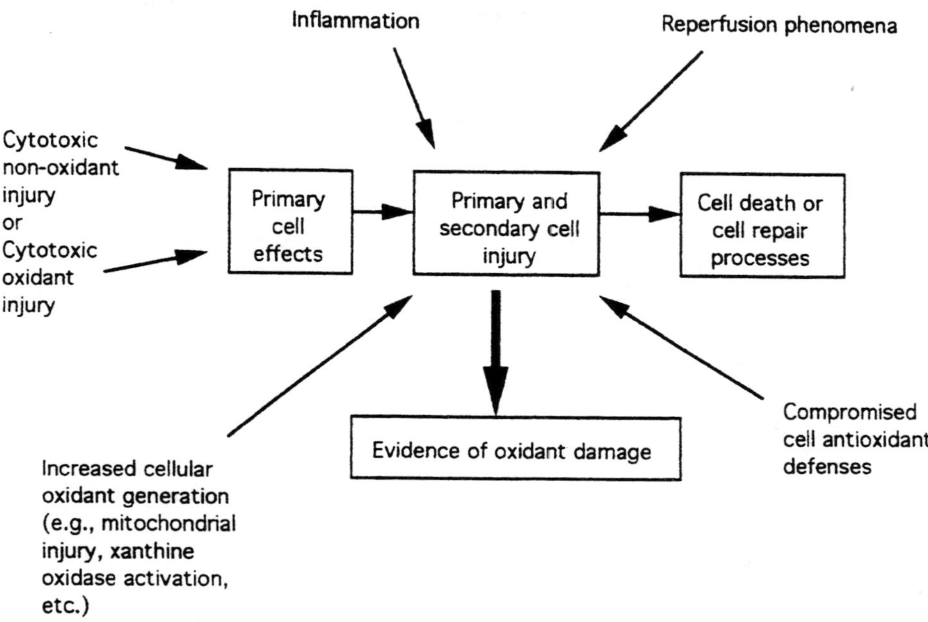

Figure 2.5. Schamatic sequence of primary and secondary cell injury, illustrating the complex, interacting nature of these processes.

Phagocytic cells also secrete another toxic compound, leukotoxin (9,10-epoxy-12-octadecenoate), which has a role in the production of acute edema.

The discussion in this section demonstrates that oxygen activation is commonly the result of metabolic processes. An increase in oxygen activation may exceed the protective activity of antioxidative systems and become an aggravating factor in several pathological conditions. This effect can occur through exposure to chemical pollutants.

2.3.3 MODIFICATIONS AT THE BODY LEVEL

As described above, oxygen activation is not usually an etiologic factor, because the body uses this phenomenon for beneficial purposes (prostaglandin synthesis) or as a defensive weapon (phagocytosis) [75, 221]. Consequently, the involvement of oxygen activation in etiology of various diseases is difficult to prove, but its implication in the aggravation of pathological conditions may be demonstrated indirectly by the nonspecific effect of some drugs or exotic treatments on the mechanism of oxygen activation, especially anti-inflammatory drugs or plants with a rich content of antioxidants (garlic or onion).

A listing of pathologic conditions where the formation of reactive oxygen species or free radicals is proven is given in table 2.5. Although there are some missing diseases, such as diabetes and pancreatitis, in which the involvement of free radicals and reactive oxygen species is still uncertain, the list is impressive in size. Toxic or polluting factors may be involved in half of these diseases. In many pathological conditions the etiology is not clearly known (i.e., cancer), but the involvement of reactive oxygen species is supported by several experimental and therapeutic factors. Based on the involvement of free radicals or reactive oxygen species, new lines of treatment have been designed for cancer, heart attack, organ transplant [34], cataract, rheumatoid arthritis, burns, and radiation exposure [56, 108, 211, 297, 300].

In agreement with *in vitro* experiments, oxygen activation *in vivo* may be increased by any supplementary source of free radicals or oxygen reactive species. According to Cross [53], the main sources of free radicals in organisms are those presented in table 2.6. These sources may be increased or affected by some physiologic or pathologic conditions, and may aggravate the progress of a disease or reduce the resistance of the body. This amplification may explain the way some chronic conditions become acute following an exposure to a xenobiotic or an infectious disease. Trauma to a tissue, including exposure to xenobiotics and inflammation, results in the release of iron through cell lysis or by the action of calcium-dependent proteases. This released iron can act as a catalyst for lipid peroxidation and membrane damage.

The mechanism of aging is a subject that has generated great interest. The hypothesis of D. Harman of the University of Nebraska, Omaha, published in 1956, is still widely accepted by scientists. The hypothesis suggests that aging is accompanied by a decreasing efficiency of antioxidant protective mechanisms and an increase in free radical formation. This suggests that both

Table 2.5
Clinical conditions in which oxygen radicals are thought to be involved
Multiple organ involvement:
        Inflammatory-immune injury
                Glomerulonephritis (idiopathic, membraneous)
                Vasculitis (hepatitis B, drugs)
        Ischemia-reperfusion
        Drugs and toxin-induced reactions
        Iron overload
                Idiopathic hemochromatosis
                Dietary iron overload (red wine, beer brewed in iron pots)
                Thalassemia and other chronic anemias
        Nutritional deficiencies
                Kwashiorkor
                Vitamin E deficiency (malnutrition)
        Alcohol
        Ionizing radiation
        Aging
                Premature aging disorders
                Immune deficiency of age
        Cancer
        Amyloid diseases
Primary (single) organ involvement
        Erythrocytes
                Phenyldrazine
                Primaquine
                Lead poisoning
                Photo-oxidation in protoporpheria
                Malaria
                Sickle cell anemia
                Favism
                Fanconi anemia
        Lung
                Cigarette smoke effects
                Emphysema
                Hyperoxia
                Bronchopulmonary dysplasia
                Oxidant pollutants (mineral dust, asbestos)
                Acute respiratory distress syndrome
                Mineral dust pneumoconiosis
                Bleomycin toxicity
                Paraquat toxicity
        Heart and cardiovascular system
                Alcohol induced cardiomyopathy
                Keshan's disease (selenium deficiency)
                Atherosclerosis

Table 2.5 continued

        Doxorubicin toxicity

Kidney

        Nephrotic antiglomerular basement membrane disease
        Aminoglycoside nephrotoxicity
        Heavy metal nephrotoxicity
        Renal graft rejection

Gastrointestinal tract

        Endotoxin liver injury
        Carbon tetrachloride liver injury
        Diabetogenic action of alloxan
        Free fatty acid-induced pancreatitis
        Nonsteroidal antiinflammatory drug-induced lesions

Joints

        Rheumatoid arthritis

Brain

        Hyperbaric oxygen
        Neurotoxins
        Senile dementia
        Parkinson's disease
        Hypertensive cerebrovascular injury (cerebral trauma)
        Neuronal ceroid lipofuscinosis
        Allergic encephalomyelitis & other demyelating diseases
        Ataxia-telangiectasia syndrome
        Exacerbation of traumatic injury
        Aluminum and mercury overload
        A-beta-lipoproteinemia

Eye

        Cataract
        Ocular hemorrhage
        Degenerative renal damage
        Retinopathies (diabetes)
        Photic retinopathy

Skin

        Solar radiation and ionizing radiation erythmia
        Burns
        Porphyria
        Contact dermatitis (including chemically induced)
        Photosensitive dyes
        Bloom syndrome

Table 2.6
Major sources of free radicals in the organism. After Cross [53]

ENDOGENOUS SOURCES
- Mitochondrial electron transport chain
- Endoplasmic reticulum electron transport chain
- Oxidative enzymes:
Xanthine oxidase
Monoamine oxidase
Lipoxygenase
Cyclooxygenase
- Phagocytizing cells:
Neutrophiles
Macrophages
Eosinophiles
EXOGENOUS SOURCES
- Compounds undergoing redox cycles:
Alloxan
Paraquat
Anthracycline antibiotics (cytostatic)
- Oxidation of chemical pollutants:
Aromatics
Solvents
- Compounds that deplete GSH:
Nitrates
Hydrazines
Sulfate
Other irritating gases
- Tobacco smoke
- Ionizing and UV radiation

aging and carcinogenesis involve free radicals, but different cells and intracellular sites are involved. For example, postmitotic, differentiated cells, such as blood vessels and brain cells show clear manifestations of aging, while stem cells appear to be more susceptible to carcinogenesis [111, 139, 183].

In both situations, reactive oxygen species produced by radiation, chemical pollutants, etc. weaken the resistance of the body. As mentioned, both ionizing radiation and the metabolism of chemical pollutants produce the hydroxyl radical ($OH^{\bullet}$). Reaction of this radical with DNA accounts for most of the DNA strand breaks caused by radiation, which, in turn, accounts for the majority of ionizing radiation-induced cancers. Oxidant stress, like heat stress, induces functional changes in gene expression [3, 121], and reactive oxygen species produced by activated polymorphonuclear leukocytes can induce DNA changes in neighboring cells. Prevention of cigarette tar-induced single strand DNA breaks by SOD, catalase, and OH$^\bullet$ scavengers (i.e., mannitol) support the idea that reactive oxygen species are involved in DNA damage.

Recently, new experimental proof has demonstrated the direct involvement of OH$^\bullet$ in DNA strand breaks [6, 74, 133]. Identical strand breaks were obtained by several means:

a) Incubation of bacteriophage DNA with buffered extracts of tobacco smoke;

b) Irradiation of DNA from various sources with ionizing radiation;

c) Incubation of DNA with a source of OH$^\bullet$ (Fenton's reaction);

d) DNA damage is reduced by the presence of free radical scavengers (mannitol) or iron chelators (desferroxiamine).

In these *in vitro* experiments, a dose-effect relationship was observed.

Strand breaks that block DNA replication mainly involve the phosphate groups at the 3' terminus. Both ligase and polymerase I possess a recognition site for that position. The blocking or altering of this position on the DNA molecule inhibits the enzymatic repair of the break. The products of *in vivo* oxidative damage to DNA include thymine glycol, 5-hydroxymethyluracil, and 8-hydroxyguanine. These products have been found in the urine of primates and normal humans, demonstrating the importance of DNA as a target of reactive oxygen species. As will be discussed later, any event that increases the production of OH$^\bullet$ will also increase DNA damage.

As observed by Ames [6, 7], cumulative risk for certain diseases increases approximately at the fourth power of age, independent of life span. Thus, it is observed that rats and mice (2 - 3 year life span) and humans (85 year life span) all have an approximately 30% incidence of cancer at the end of their life spans. This suggests one important factor in longevity may be basal metabolic rate, which is much lower in humans than rodents. The lower metabolic rate may lower the production of endogenous mutagens during normal metabolism. Rats have a high specific metabolic rate (SMR) and are observed to eliminate more thymine glycol. A linear relationship can be demonstrated between SMR and the elimination of DNA oxidative stress compounds (figure 2.6). The SMR of the organism is dependent on the rate of O_2 utilization of tissues, making it proportional to the rate of reactive oxygen species formation. Therefore, the ratio of total antioxidant concentration (enzymatic and nonenzymatic) to SMR indicates the degree of protection from reactive oxygen species.

For mankind, the Median Life Span Potential (MLSP) is considerably greater than in other species. Aging seems to be a product of both SMR and environmental factors. Cutler [56] observed that the rate of aging is proportional to SMR, and for mammals the product of MLSP and SMR is a constant. From this he derived the relation:

$$LEP = 2.70 \, MLSP \times SMR$$

where LEP is Life Energy Potential, expressed as kcal/kg. If MLSP is proportional to the ratio of antioxidants to SMR, the LEP value of a species is proportional to the total concentration of antioxidants. For most species of mammals, LEP falls between 200 - 300 kcal/kg, but for man the LEP is 833 kcal/kg [34, 108]. A close correlation has been found between SOD content of liver and MLSP (r = 0.727) and between SOD/SMR and MLSP (r = 0.961, p<0.01).

The last proof for the involvement of oxygen activation in aging is provided by the lipofuscin stored in various cells, mostly those with a high metabolic rate (heart and brain). Lipofuscin pigments are the single pathological modification produced in senescence. The structure of lipofuscin pigments is complex and is dependent on the type of cell involved. Lipofuscins contain variable amounts of unsaturated phospholipids, peroxidized PUFAs, phosphatidylcholine, phosphatidylserine, and large amounts of copper and iron. Lipofuscin pigments are believed to result from the action of reactive oxygen species on proteins, nucleic acids, and PUFAs from the endoplasmic reticulum. Lipofuscin granules have dimensions ranging from 0.5 - 3.0 μ and are fluorescent. The amount of lipofuscin increases under conditions that favor peroxidation and the release of proteases, such as intoxication, stress, and exertion.

 R. Olinescu, T. Smith and J. Hertoghe

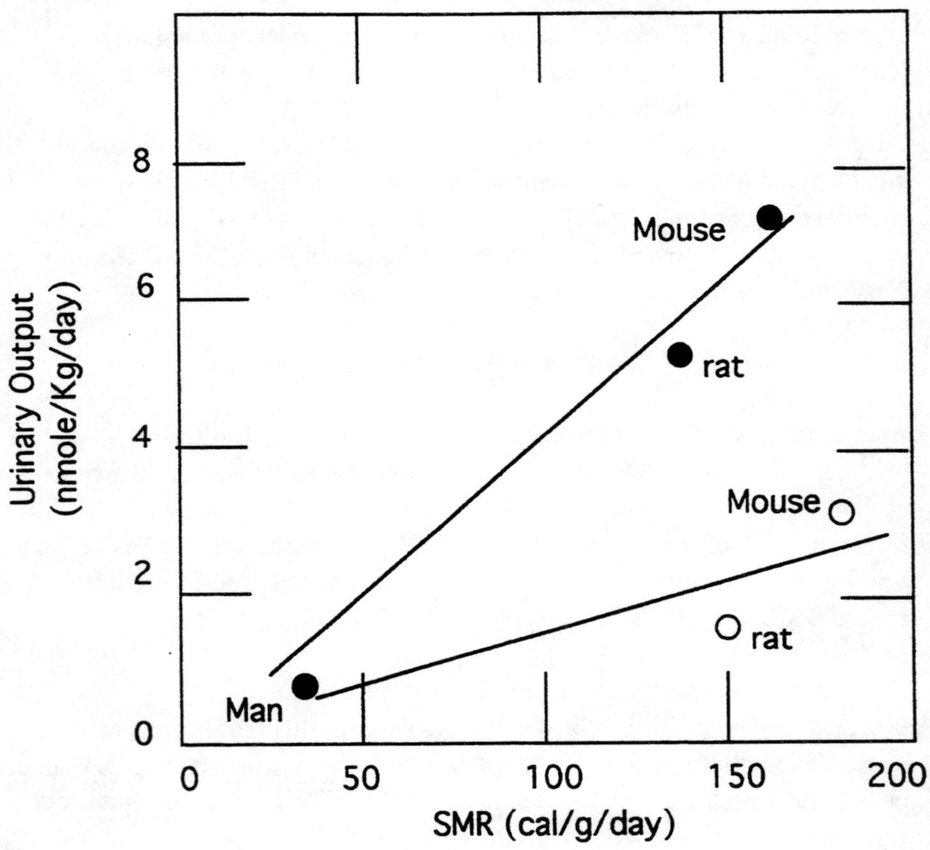

Figure 2.6. Average urinary output of thymine glycol (closed circles) and thymidine glycol (open circles) for three species, expressed as a function of the specific metabolic rate (SMR) of that species. Urine levels are mean values± SE for untreated humans, rats, and mice. From Ames [6,7].

PART II: THE BODY'S DEFENSES

CHAPTER 3

ENTRY OF CHEMICAL POLLUTANTS AND MOVEMENT INSIDE THE BODY

3.1 SOURCES OF POLLUTANTS

A brief summary of the sources of pollution includes:

a) Combustion in domestic and industrial settings. Combustion forms carbon monoxide, carbon dioxide, sulfur oxides, partial combustion products (soot), polycyclic aromatic hydrocarbons, and other chemicals.

b) Industrial discharges of manufacturing byproducts: asbestos, polychlorinated biphenols (PCB), solvents, radioactive waste, etc.

c) Exhaust from various means of transportation (automobiles, airplanes, ships, and trains) using fuel refined from oil. Waste products include incomplete products of combustion, fine particles of lead from tetraethyl or tetramethyl lead fuel additives (where their use is still allowed).

d) Chemicals used in agriculture: pesticides, herbicides, animal-feed additives, and fertilizers.

e) An ever increasing number of household products including, cleaning solvents, detergents, paints and varnishes, laundry and dish-washing products, insecticides, cosmetics, plastic packaging materials.

f) Both added and naturally occurring chemicals in food.

Food is the most chemically complex substance encountered by the

Table 3.1
Mean daily intakes of natural compounds in the U.K.
(after Morgan and Fenwick [193])

Class of Compounds (source)	Ingested (mg/day) by:	
	Total population	Vegetarians
Glucosinolates (brassicans)	50	110
Glycoalkaloids (potatoes)	13	70 - 90
Saponins (vegetables)	15	100 - 200
Isoflavones (soya)	<1	105

public. There are probably more than half a million naturally occurring compounds in fresh plant foods and more are formed by preservation and cooking. There are also fewer than 1,000 approved agricultural chemicals and food additives. Even food in its natural state should not automatically be considered free of toxic compounds. The amount of some compounds that can be obtained from dietary sources is shown in table 3.1. Not all of these should be considered toxic.

Increased incidence of goiter and cretinism has been associated with a diet high in cassava and other cyanide containing plants. Cabbage and brussels sprouts contain a potent goiterogen, 5-vinyloxazolidine-2-thione. Legumes contain a wide range of biologically active compounds, such as estrogenic isofavones, saponins, hemagglutinins, phytates, polyphenols (tannins), enzyme inhibitors, etc.

Fish, especially shellfish, can contain very potent toxins such as demoic acid in mussels or histamine in fish. Mycotoxins produced from certain fungi, notably aflatoxins, can also be present in shellfish and plant materials.

These sources and others contribute to a widespread and complex exposure pattern to potentially harmful chemicals in the environment. The complexity stems, in part, from the many factors that can alter the consequence of an exposure (physical and chemical form of the agent, mode of entry into the body, dose, etc.). Pollutants carried by the air are very important, not only directly as gases and vapors, but for solids of a sufficiently small size to produce an aerosol as well. These particles may be toxic on their own or they may be carriers of toxic materials. When aerosol particles settle or are washed out of the air by rain, they can contaminate the upper layers of the soil.

The effect of exposure to chemical pollutants depends on the route of

Table 3.2
Chemical pollutants and the body

A. Exposure	-chronic -acute
B. Routes of entry	-respiratory system -digestive system -skin
C. Transformation	-transport (blood) -distribution in tissues -metabolism
D. Storage (variable period)	-tissues
E. Elimination	-respiratory system -bilinary system (via intestines) -skin

entry and what happens once they are inside the body. The chemical structure of pollutants determines the subsequent events in the body. Some pollutants produce adverse effects immediately, while others have no effect for years. Some are eliminated after a few hours; others are stored for long periods of time. A generalized sequence of events in chemical exposure is given in table 3.2, but the steps and timing are frequently unpredictable. This is at variance with pure toxins, poisons, and drugs, which are the topic of most books on toxicology [63, 122, 147, 298]. The second factor, route of exposure, adds to the uncertainty surrounding the action of chemical pollutants, since multiple exposure routes are common.

A third factor is the effect of long-term continuous absorption of low concentrations of pollutants. These concentrations are too low to produce clinical symptoms of toxicity. Pollutants that exhibit a cumulative effect include the heavy metals (Hg, Pb, Cd, etc.) and insecticides (DDT).

Synergistic and additive effects of chemical pollutants have been reported. Thus, the toxic effect of one compound at low concentration can be increased by exposure to a second toxic compound. For example, exposure to herbicides can increase the toxic effect of exposure to insecticides. Synergistic effects are also known for heavy metals. The presence of some types of bacteria may enhance the absorption and toxicity of some pollutants, as occurs in the formation of the organic mercury compound, methylmercury. Under certain conditions, pollutants can undergo transformation into new, possibly toxic, compounds. For example, automobile and aircraft engine exhaust form irritants and toxic peracyl-nitrates through photochemical reactions. These are major constituents of the "oxidizing smog" found in some cities, such as the Los Angeles area [119, 207, 272].

There is always the possibility of reactions between various chemicals present in a given environment. This has raised the question of the possibility of the formation of toxic organochlorine compounds when water containing large numbers of bacteria is treated with chlorine. Water containing traces of phenolic compounds, which have no flavor, can react with chlorine to produce foul tasting chlorophenols.

A final point to be made here is that most pollutants are eliminated from the body at some time after they enter. Figure 3.1 is a more detailed representation of the information presented in table 3.2 and illustrates the movement of pollutants through the body and their eventual elimination.

3.2 ROUTES OF EXPOSURE

As is true in classical toxicology, chemical pollutants exert toxic effects depending on the mode of entry into the body. The mode of entry can affect the rate and amount of a pollutant that enters the body. The faster a pollutant is absorbed and the more that gets into the system, the greater the toxic effect of the chemical. In an industrial environment, the respiratory route is a major means of entry of contaminants, accounting for 80 - 90% of all pollution exposure [63, 207, 247].

The absorption of chemical pollutants proceeds according to the laws of chemical kinetics, as is true for drugs and other toxic substances (table 3.3). In absorption following a zero order kinetic, absorption occurs at a constant rate, depending on the compound's concentration. The more the concentration of the chemical, the more is absorbed. Inhalation of toxic gases (such as sulfur oxides), absorption through the skin, and the mobilization of substances out of subcutaneous and intramuscular sites and from various organs (such as the liver) are examples of events following zero order kinetics.

In absorption by a first order kinetic rate, the amount absorbed is constant for each unit of time, regardless of the amount of chemical present. This is the common kinetic for absorption of materials from the digestive tract.

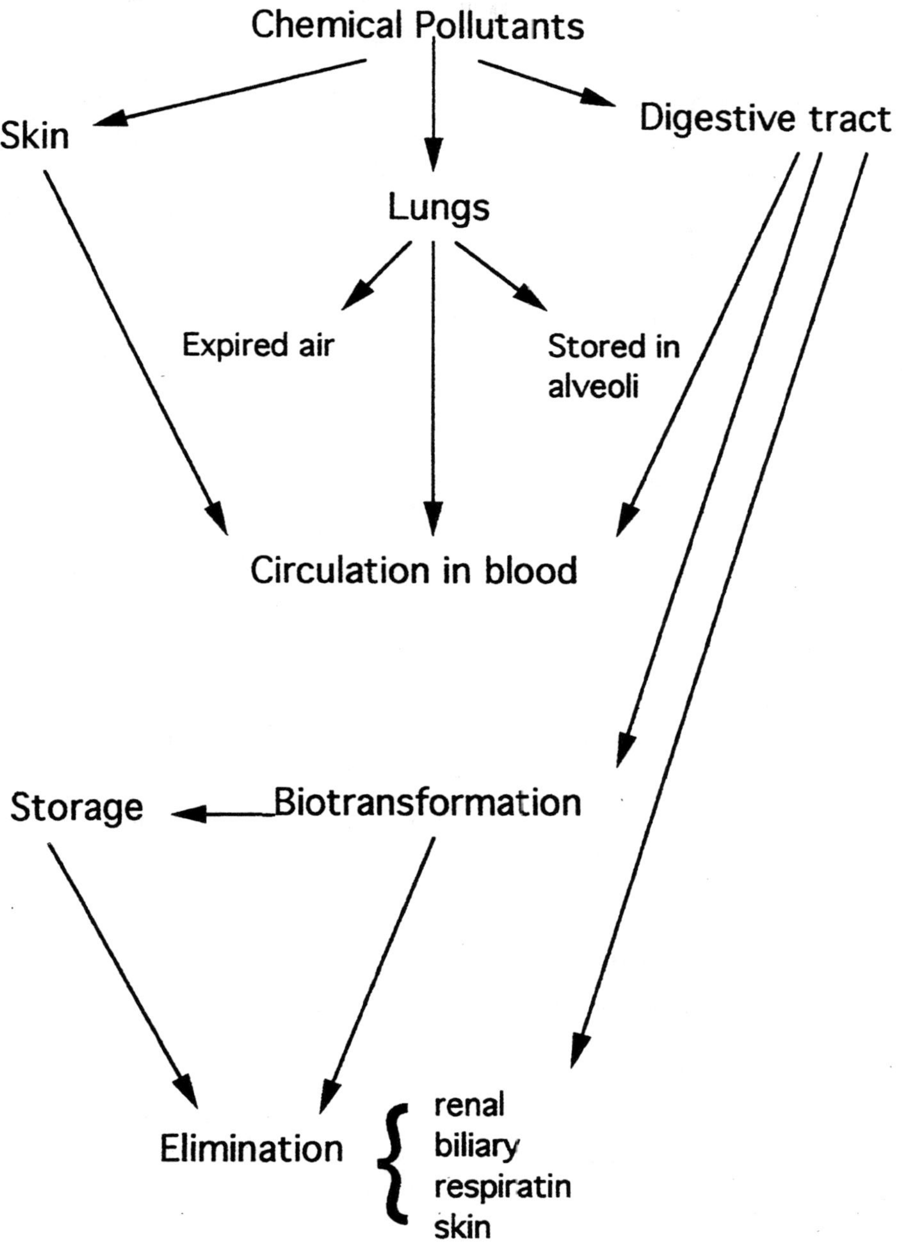

Figure 3.1. Movement of pollutants in the body.

Table 3.3
Characteristics of absorption rates

Order	Equation	Unit of rate Constant (K)	Characteristics
Zero	$d[x]/dt = K°$	$mol \cdot s^{-1}$	Rate is constant and independent of the chemical's concentration
First	$d[x]/dt = K'(a-x)$ where a = the initial concentration of x	s^{-1}	For each unit of time, the same amount of chemical is absorbed

3.2.1 RESPIRATORY EXPOSURE

The respiratory tract is a common means of exposure to pollutants. The absorption of a substance by this route is influenced by the physical and chemical structure of the compound. The effect of absorbed material will be further altered by individual variation in threshold sensitivity and by adaptation. The large surface area of the lungs (over 100 m^2), with its intimate association with the blood circulation makes the lungs a means of entry into the body for a variety of chemical pollutants, including gases, aerosols, and particulates (dust, smoke, fumes). The high vapor tension of some substances in the gaseous or liquid states insures an efficient absorption by this route [17, 58].

During evolution, nature provided the respiratory tract with mechanical obstacles to penetration by dust and fumes. Consequently, the anatomical structure limits the size of dust and particulates that can get to the lungs. The nose and pharynx contain hair and a mucous layer, which traps dust, allowing its later elimination by sneezing, etc. Particulates larger than 10 microns in diameter are screened out in the nose and pharynx.

The trachea and bronchial tubes contain seromucous glands, and are covered with a layer of mucous and motile cilia. This section of the respiratory tract traps particulates that range in size between 3 and 10 microns in diameter. The lungs contain about 400 billion alveoli, grouped in clusters of 20. The alveoli are coated with a surfactant, which is a complex mixture of lipids, proteins, and other substances that act to reduce the surface tension at the surface of the cells. Gases are dissolved in the surfactant before being transferred to the blood. The smallest particulates also become trapped by the surfactant [17, 58].

Particulates are collected by the mucous layer and swept out of the respiratory tract by the movements of the cilia. This movement amounts to an estimated 1000 pulses per minute, allowing a 14 - 18 mm per minute movement of particles [247]. Coughing completes the removal of particles trapped by this defensive mechanism. This mechanism allows the adaptation of man to living in dusty regions, such as deserts.

An adult male inhales about 6 L of air, containing about 250 ml of oxygen per minute and eliminates by expiration about 200 ml of carbon dioxide. During physical effort, these amounts can be increased by 10 - 20 fold. This means an impressive figure of 20 - 30 thousand liters of air, containing dust, gases, and chemical vapors passes through the lungs daily. The volatility and lipid solubility properties of gases determine their penetration in the lungs. Some non-reactive gases, such as some organic solvents are reduced in their penetration. In contrast, very reactive acids and bases can produce significant injury [256, 287].

Metallic dust and particulates in aerosols are a separate case. Many of these particles have an electric charge, preventing them from aggregating. Their small size allows them to penetrate deep into the lungs. The effects of exposure to pollutants by the respiratory route will be covered in more detail in Chapter 8.

3.2.2 GASTROINTESTINAL EXPOSURE

The second major means for chemical pollutants to enter the body is by the gastrointestinal system. As in the respiratory system, the gastrointestinal system has a large surface area to enhance the absorption of desired nutrients. The length of the intestines allows selective absorption of nutrients by different parts of the intestine and extends the time during which absorption can occur. While some pollutants can be absorbed in the mouth, most will be absorbed in the gut, where the large surface area and vascularization favors this. As is the case for drugs, the absorption of pollutants is limited by their solubility in the acidic pH of the stomach and first few centimeters of the small intestine.

Chemical pollutants may cross biological membranes, especially those of the gastrointestinal tract, by passive diffusion, or by active transport. Some chemical pollutants, with structures similar to nutrients, show saturation in their absorption rate at higher concentrations. This provides evidence they are being taken up from the intestine by active transport [8, 290].

Lipid solubility is the main factor governing the absorption of organic compounds. For ionic and polar compounds, their degree of ionization and the lipid solubility of the non-dissociated form is of primary importance [63, 143].

Lipid soluble chemical pollutants cross the gastrointestinal mucosa by passive diffusion. Small polar compounds, such as ethanol, or water, cross by passive diffusion, passing through the intracellular spaces. For organic pollutants, such as purines, aromatic hydrocarbons, or macromolecules, active transport, requiring an energy supply, is most likely involved [24, 49].

Organic acids are absorbed mostly in the stomach, since at an acidic pH they are non dissociated. Weak acids and bases are found in the small intestine, especially the duodenum. The colon is the primary site for the absorption of water and electrolytes.

3.2.3 TRANSCUTANEOUS EXPOSURE

The skin has a surface area of approximately 1.5 m^2, but the most exposed regions (the face and hands) amount to only 0.25 m^2. The skin has three main layers that provide protection from external agents. At the same time the skin is able to provide a slow but continuous absorption, especially when compounds are present as residues or creams. [65, 287].

When chemical pollutants are able to dissolve in the skin's tissues, a concentration gradient forms across the skin. Lipid soluble compounds easily pass through the skin, but water soluble compounds must first dissolve in the sebum, then diffuse inward to the blood vessels. Absorption by skin is accelerated by rubbing, perspiration, and small scratches or scrapes [63].

Some gases, such as H_2S and CO, and a large number of organic compounds (hydrocarbons, aromatic derivatives, insecticides), and possibly metals (Pb, Bi, Hg) can pass across the skin. The presence of detergent on the skin increases the ability of compounds to cross the skin.

Absorption of chemical pollutants through the eyes is rapid and potentially very dangerous. Various gases, alkaloids, and insecticides can easily pass across the conjunctiva and cornea.

3.3 TRANSPORT, DISTRIBUTION AND TRANSMEMBRANE MOVEMENT OF CHEMICALS

After entering the blood, the distribution of chemical pollutants in the body varies, and is less well understood than their absorption, metabolism, and excretion. In the body, chemical pollutants may pass through several steps of transportation and transformation, as summarized on table 3.2 and figure 3.1.

Acute intoxication or drug administration results in a defined and predictable level of the agent in the tissues. Exposure to pollutants is generally less predictable, resulting in plasma concentrations that are less predictable and subject to cumulative effects.

As shown in figure 3.2, the plasma concentration following exposure to chemical pollutants gives only limited information on overall exposure. The immediate plasma level depends on a variety of factors, including the timing of exposure, limits of detection, and rate of metabolism. It is only for a continuous or acute exposure that pharmacodynamic or toxicologic relationships apply.

3.3.1 BLOOD TRANSPORT

Like the majority of foreign compounds as well as for naturally occurring compounds (i.e., bilirubin, hormones, etc.) chemical pollutants reversibly bind

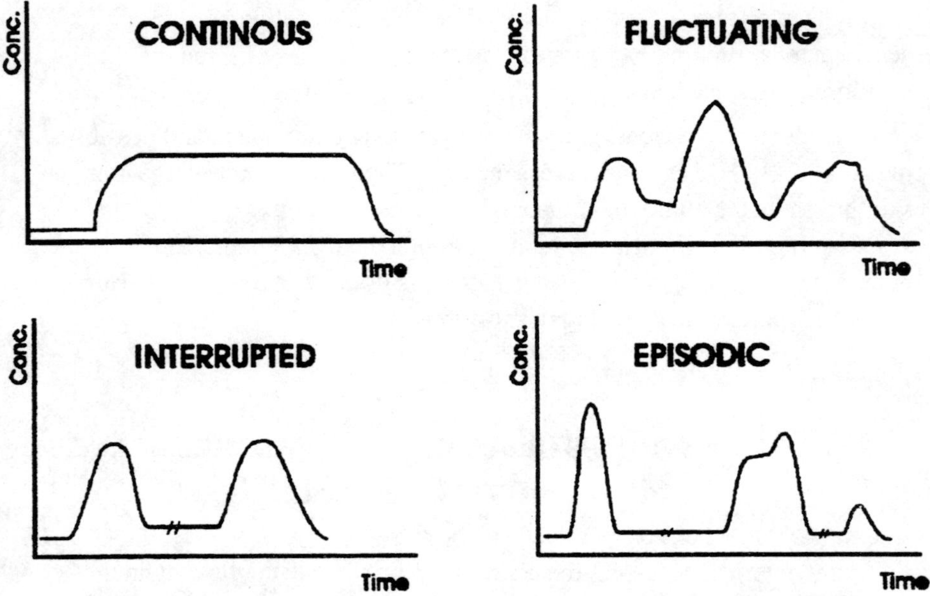

Figure 3.2. Variations in blood concentration of an agent following its penetration into the body. After Voicu and Olinescu [287].

to plasma proteins or to erythrocyte membranes. The later is not well known and has been conclusively proved only for lead. But the plasma proteins, especially albumin, bind and transport a large variety of compounds, such as solvents (benzene), antibiotics (penicillin), drugs, vitamins, fatty acids, metallic ions (Ca^{2+}, Zn^{2+}, Cu^{2+}), and metabolic products (bilirubin, uric acid).

The reversible binding of chemical pollutants to plasma proteins utilizes polar and Van der Waals bonds. In addition to hydrophobicity, the binding of chemical pollutants to plasma proteins is influenced by pH, as an acidic environment favors binding. The relationship between free compounds (X), and protein bound (PX), with n active sites, and dissociation constant K is:

$$[PX]/nP = [X]/(K+[X])$$

By defining the amount of chemical pollutant bound to a protein molecule as r, the equation can be rearranged:

$$r = [PX]/P = n[X]/(K+[X])$$

The r vs. [X] relationship gives a curve identical to the absorption curve resulting from saturation at high concentrations [287]. In the case of a chemical pollutant with a high binding affinity (low K) all will be bound, even at a low concentration. At sufficiently high concentrations, any compound can saturate the binding sites on plasma proteins, resulting in a significant amount of free compound, possibly increasing the metabolic rate. As the total concentration of the chemical pollutant decreases through metabolism and excretion, the bound fraction increases in relation to the total concentration [58, 76].

Chemical binding to plasma proteins tends to trap the agent in the blood, delaying its movement across the cell membrane. When binding sites on albumin are saturated, there is competition for binding between the chemical pollutant and natural compounds normally carried by albumin in the circulation. It is possible to increase the transport capacity of the blood by using artificial carriers, such as polyvinylpyrolidone, dextran, or complexing agents, such as EDTA and dimercaptopropanol. The potential for these interactions should be taken in account for people with diabetes or liver disease who are exposed to a toxic environment.

3.3.2 DISTRIBUTION

Distribution is defined here as the overall movement of chemical pollutants from the blood to tissues. Passing through the walls of the capillaries, chemicals enter the interstitial fluid. The walls of capillary vessels behave like a selectively permeable membrane. Lipid soluble substances pass the capillary wall over the entire surface at a rate depending on the molecule's lipid/water partition coefficient. Water and electrolytes, as well as molecules with a molecular mass of less than 400 traverse the capillary wall by diffusion. The disappearance of chemical pollutants from the blood can occur following first order kinetics, as shown in figure 3.3.

Portion "a" of the curve shown in figure 3.3B represents the blood concentration of a chemical pollutant when there has been acute exposure, and the compound is being moved into the tissues. For free chemicals, the plasma concentration is proportional to the tissue concentration. The middle portion of the curve (part b) corresponds to metabolism of the compound, or excretion of a nonmetabolizable compound. On a semilog plot, this part of the curve becomes straight and can be used to determine the half-life ($T_{1/2}$) of the compound. Part "c" indicates excretion.

The distribution of a chemical pollutant may follow one of several alternatives:

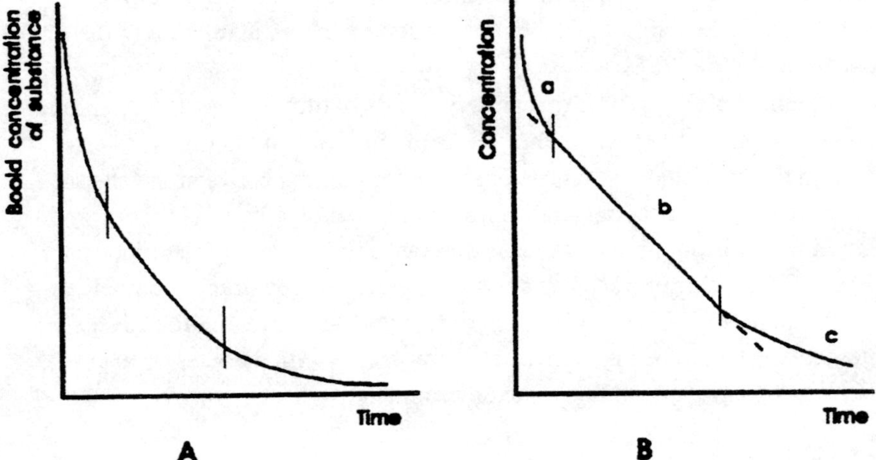

Figure 3.3.   The decrease in blood concentration of a xenobiotic following acute exposure.

- Uniform distribution in all tissues (e.g., alcohol, narcotics, etc.)
- Selective distribution to certain tissues due to differences in the permeability of cell membranes. For example, iodine accumulates in the thyroid, mercury salts in the kidneys, calcium and strontium in the bones, and cesium in the muscles.
- A general distribution, in which an agent is metabolized in the liver and excreted or stored. A substance that reaches the liver can be excreted in the bile without transformation, from which it is eliminated in the feces or reabsorbed through the intestinal wall, forming an enterohepatic cycle [63].

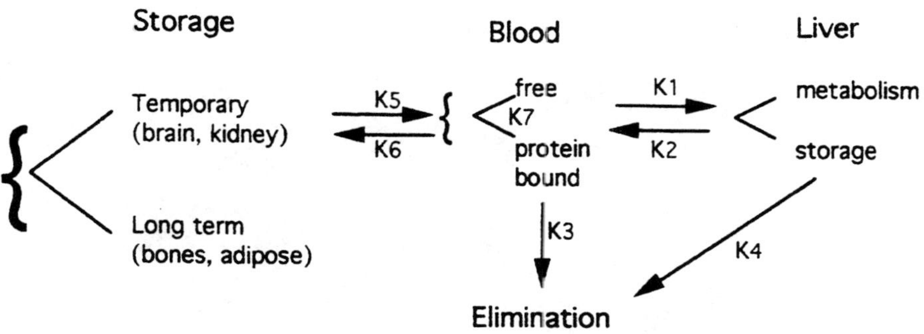

Figure 3.4.   Distribution of chemical pollutants in the body.

Most chemical pollutants are distributed through the body following one of the routes indicated in figure 3.4. Because there is great variation in the structure of chemical pollutants, there is also variation in the distribution and elimination of these compounds. Because of the law of mass action and binding competition, chemical pollutants and their metabolites will be moved through the blood towards organs and tissues. As determined by their structure, the rate of distribution can be partially predicted:

K1>K3>K4 for alcohol, and other volatile substances with rapid metabolism.

K7>K5>K6>K3 for substances with rapid metabolism, and for metallic ions with selective distribution.

K7>K1>K2>K4>K5 for most organic compounds.
In this figure, bile excretion and reabsorption are not shown. Many chemical pollutants that do follow this path produce biliary cholestasis (blockage of the bile duct).

Storage in tissues is reversible, temporary, or long-term, and may involve binding to lipoprotein structures of cell surface receptors. By binding to these sites, chemical pollutants may interfere with the normal biologic response to hormones, vitamins, or drugs. Competition at the level of receptors depends on a similarity of structure and the binding strength. The binding energy for a

insert figure 3.5

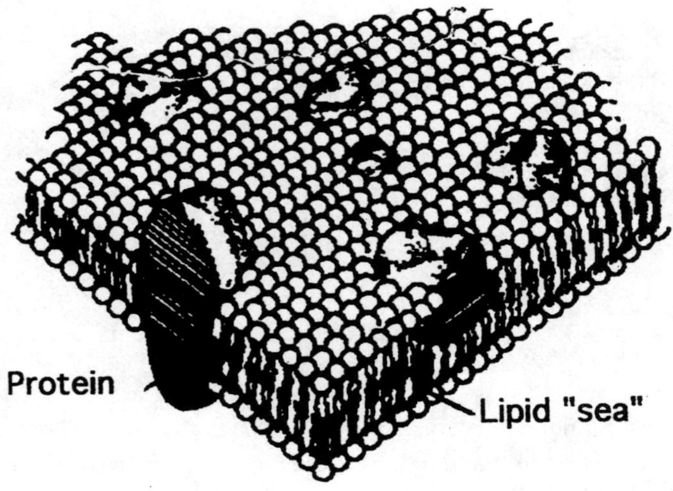

Figure 3.5.   The Signer-Nicholson model of cellular membrane structure.

receptor-compound complex may vary from 20 - 100 kcal/mol (covalent bond) to 5 kcal/mol (hydrogen bond). There is no widely accepted theory of the interaction between xenobiotics and protein receptors, even though this has been extensively studied.

3.3.3 TRANSMEMBRANE MOVEMENT

Any compound that enters the body or is synthesized in the body must cross a variable number of cellular membranes and barriers (such as the blood-brain barrier). The blood-brain and placental barriers represent a category that differs from the cell membrane. In the case of these barriers, chemical pollutants pass by diffusion, determined by molecular size and selectivity.

According to the Singer-Nicholson model, biological membranes possess a dynamic structure, consisting of a phospholipid bilayer traversed by proteins, which provide hydrophilic zones (fig. 3.5). Such a structure is not thermodynamically stable, but is maintained by the inward pressure of the aqueous free energy and by polar protein and carbohydrate groups in contact with the aqueous phase. Membrane structure is not rigid, but exists in a complex state of continuous flux.

This membrane structure favors the penetration of lipid soluble substances, which includes 70% of chemical pollutants. The membrane structure includes protein channels, which provide for selective transmembrane movement of electrolytes. These channels, having diameters in the order of 5 Å can be opened in response to a stimulus, such as a chemical messenger like cAMP or variations in electrical potential.

Transport of molecules and ions across cell membranes involves the mechanisms of:

a) *Passive diffusion*, occurring because of a chemical concentration gradient across the membrane. Fick's law defines the diffusion rate as KA {(c1 - c2)/d}, where A is the surface area available for diffusion, (c1 - c2) is the concentration gradient, d is the thickness of the membrane, and K is the diffusion constant, which is a function of the molecular mass, configuration, and ionization of the substance. The undissociated form of many compounds is more lipid soluble and diffuses more easily than the dissociated form. Low molecular weight compounds commonly enter cells by this means [76].

b) *Active transport*, which acts in opposition to the concentration gradient and, consequently, requires an energy source. This energy derives from the energy producing pathways (glycolysis, Krebs cycle, and electron transport chain) through the energy storage function of ATP, and is released by the hydrolysis of the ATP.

c) *Mediated transport* requires the presence of a carrier protein in the membrane. Most nutrients (glucose, amino acids) use this form of transport. Mediated transport follows first order kinetics in its rate, and is subject to saturation, steriospecificity, and competitive and noncompetitive inhibition (involving the SH and NH_2 groups on

proteins). The rate of mediated transport is limited by the number of carriers available, and by the speed with which the substrate can be bound and released.

d)   *Ion pumps* are used for movement of sodium and potassium ions, involving ion channels through the membrane. Ion pumps require the use of energy released by the hydrolysis of ATP, which in turn is dependent on the enzyme ATP-ase. The activation of this enzyme depends on an adequate concentration of relevant ions on both sides of the membrane. The activity of this enzyme can also be increased by certain drugs, such as phenothiazines and cardiac glycosides.

Polluting ions (Cd^{2+}, Zn^{2+}, Li^+) with similar size and charge to potassium and calcium can interfere with ion pumps. It is also unfortunate that, while the movement of nutrients into the cell requires an energy expenditure, many chemical pollutants have sufficient solubility in the lipid membrane (halogens; nitrogen, sulfur, and phosphorous derivatives; solvents; aromatic compounds) that they pass by simple diffusion [45, 67, 82].

The dynamic nature of cell membranes favors gradual and limited adaptation to chronic, low-level exposure to chemical pollutants. Such adaptation is only possible if the membrane includes an enzyme capable of detoxification of peroxides and removal of free radicals generated by the pollutant. The primary source of lipid peroxidation is the cellular lipid membrane.

A number of substances can destroy or modify the cell surface, including irritating gases (SO_2), strong acids and bases, and organic compounds.

Chemical pollutants that reach the cell surface can alter electric charges on the surface, interfering with the function of natural receptors. Through this mechanism, pollutants can alter the metabolic processes of the cell, triggering or inhibiting normal events, such as the activation of glycolysis by second messengers like cAMP. Phospholipase A2 is an important enzyme for the mobilization of fatty acids (arachidonic acid) from the membrane for use in the synthesis of prostaglandins. This enzyme can be activated or inhibited by exposure to solvent or metallic ion pollutants [85, 163].

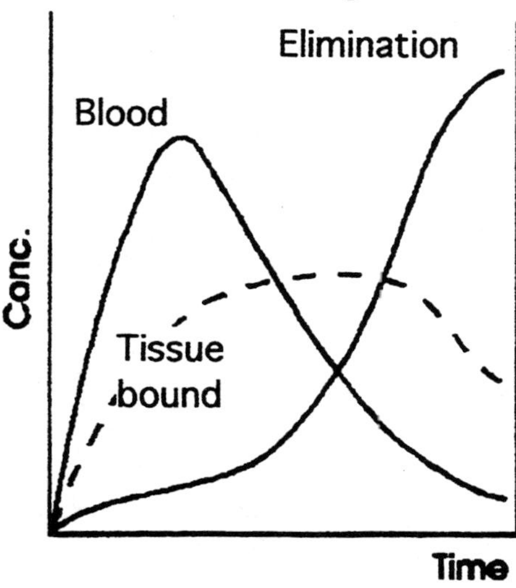

Figure 3.6. Variations in the concentration of a xenobiotic in the blood, urine, and tissues, following first order absorbtion rates.

3.4 ELIMINATION OF CHEMICAL POLLUTANTS

As indicated in figures 3.1 and 3.6, elimination is the last stage in the pharmacokinetics of chemical pollutants. Elimination from the body makes use of all possible routes: respiration, fecal excretion, and urinary excretion. A chemical pollutant, like any other xenobiotic, may be eliminated by the action of more than one route, although one route generally predominates. In general, water-soluble substances are eliminated through the kidneys, those less water soluble through the intestinal tract, and gaseous or volatile ones through the lungs.

The methods of elimination of chemical pollutants from the body may follow zero order (constant rate) or first order (exponential rate) kinetics, as illustrated in figure 3.6. Typical examples of first order events are glomerular filtration, absorption through the gastrointestinal mucosa, and sequestration of compounds in tissues. These processes follow first order reactions because each molecule of pollutant has a fixed probability of being acted on in a given unit of time. Only in the case of concentrations high enough to cause saturation does elimination seitch to zero order kinetics [20, 287].

Whether the body is eliminating chemical pollutants or drugs, the rate and route of elimination depends on concentration and chemical structure. As shown regarding absorption, contamination with a pollutant will result in the compound appearing in the blood, then moving into the tissues. The tissue levels of the contaminant will show a plateau lasting a certain period (fig. 3.6). This is Goldstein's "plateau principle," which is applicable to any system in which the input rate is constant and the elimination is exponential. The amount of the substance in the body reaches an equilibrium, as illustrated by the presence of the plateau. The duration of the plateau depends on the elimination rate and the saturation of the elimination route.

Following repeated exposure to a low concentration of a pollutant, saturation of the elimination route is reached, as shown by the nearly constant levels detected in the blood and urine. The constant level in the blood while elimination is occurring implies storage has taken place in the tissues. As mentioned in chapter 3.3, because a chemical pollutant tends to equilibrate between various tissues, the amount measured in the blood and urine provide sufficient information to determine the degree of contamination.

Elimination of pollutants through the kidneys, by glomerular filtration (particularly for small molecular weigh substances), tubular secretion, or both is the most common route. Glomular filtration yields an ultrafiltrate containing the pollutant, in the native state or as metabolites, at a concentration approximating that found in the plasma. Tubular secretion may be active or passive.

Like other biological membranes, the distal tubules act as lipoprotein barriers, preventing the transfer of non ionized, lipid soluble substances. Consequently, lipid soluble compounds remain in the blood and are distributed between the lipid and aqueous portions as determined by the compound's partition coefficient. Compounds that are more ionized in urine than blood tend to pass from the blood into the filtrate. Lipid soluble compounds that have not been metabolized into hydrophilic compounds remain in the circulation longer (as is the case for insecticides).

Compounds eliminated by active transport are strongly ionized and can be transferred to the tubular urine against a high concentration gradient. Chemical pollutants eliminated by active transport compete with each other reducing their excretion rate, as can be observed for aromatic compounds [63, 76].

Polar, lipid soluble compounds are, generally, retained in the blood to a lesser extent than undissociated lipid soluble compounds. The products of enzymatic conjugation are water soluble, strongly ionized at blood pH and, therefore, are eliminated more rapidly. The role of xenobiotic detoxification and metabolism is to increase the water solubility of the compound. Therefore, acidic or highly polar compounds are not metabolized.

Excretion in the bile occurs primarily in the case of very polar metabolites conjugated to glucuronic acid. In this way elimination occurs with the feces. Due to the high permeability of the hepatic parenchyma, the blood-bile barrier allows metabolites to pass with molecular weights less than 400. The conjugates excreted in the bile can be hydrolyzed by the intestinal microflora, sometimes resulting in products that are more toxic. As an example, aromatic polycyclic amines, when hydrolyzed, produce carcinogenic o-hydroxylamines. Normally, the bilinary elimination route involves concentrating the compound in the gall bladder. When the bile is released from the gall bladder in response to a digestive stimulus, a high concentration of the xenobiotic metabolite is put into the intestine [114, 236].

3.5 CHRONIC TOXICITY

Toxic effects can be produced by acute or chronic exposure to chemical agents. Acute exposure is defined as a single or multiple exposure occurring within a short time (24 hours or less). Chronic exposure is defined as daily or otherwise repeated exposure over a long period of time, e.g., over the working life-time or the entire life span. For many agents the toxic effects of acute exposure are quite different from those produced by chronic exposure. In general, acute exposure to agents that are rapidly absorbed produces immediate toxic effects. In some cases, however, acute exposure can result in delayed toxicity, i.e., toxic effects occurring several hours or days after exposure. Long-term, low-level exposure usually does not produce immediate toxic effects, but some indications of toxicity may become apparent on a short time scale.

Toxic effects generally occur when the agent accumulates in the body (absorption exceeds metabolism and/or excretion), or when a compound produces irreversible toxic effects, or when there is insufficient time for the body to recover from the toxic effect prior to subsequent exposure. When the elimination rate is initially less than the absorption rate, the compound does not usually accumulate indefinitely, but through adaptation and storage reaches a dynamic equilibrium, where the rate of elimination equals the rate of absorption.

An example of chronic toxic effects produced by the accumulation of a metal in a specific organ is the nephrotoxicity of inorganic cadmium and mercury ions. The initial accumulation may proceed without effect on the function of the kidneys, but, when a critical concentration is reached, damage occurs (kidney damage from mercury exposure can also be produced through the action of the immune system).

Accumulation of metals in tissues does not necessarily imply a toxic effect will occur. In the case of some metals, inactive complexes or storage molecules are formed. Lead, for example, is stored in bones and teeth in an inert form. Cadmium and some other metals bind to a cysteine-rich, low-molecular weight protein, metallothionein, which appears to be an inert storage form. It is thought that kidney damage is prevented as long as the tubular cells can produce enough metallothionein to bind the toxic cadmium ions. If this capacity is exhausted, free cadmium ions appear in the cells and are toxic.

For some metals the toxic effect is due to irreversible events, such as the action of methymercury on the central nervous system. Although all of the biological mechanisms are not yet understood, the carcinogenic activity of some metals and metallic compounds appears to also result from such irreversible events.

Toxic effects can be divided into local and systemic effects based on their site of toxic action. Local effects refer to those that occur at the site of first contact between the biological system and the toxic agent. Depending on the route and circumstances of exposure, the gastrointestinal tract, respiratory organs, skin, or eyes can be affected. Gastrointestinal effects that may occur as a result of repeated ingestion of toxic compounds over a period of time include loss of appetite, nausea, vomiting, and diarrhea followed by constipation. Intestinal spasm and pain has been observed in factory workers and children with relatively low-level lead exposure. Chronic pulmonary disorders, such as toxic and allergic lung diseases and lung cancer result from inhalation of certain chemical pollutants and metal compounds (e.g., arsenic and beryllium compounds, cadmium oxide, chromate, nickel and nickel compounds, hexachloroplatinate). Other local toxic effects include allergic skin reactions, which can be induced by many pollutants.

Systemic effects require absorption and distribution of the toxic agent to a site distant from its entry point. Chemical compounds may produce a variety of systemic effects depending on the specific site of the organism being affected. Usually, major toxic effects are limited to one or two organs. These organs are called the target organs for the toxicity of that chemical. The target organ is not always the site of the highest concentration of the agent. For example, lead is concentrated in the bones, but its chronic toxicity is mainly directed to the blood producing and nervous systems.

For some substances both local and systemic toxic effects can be observed. For example, tetraethyl lead produces effects on the skin at the site of absorption and is then transported to produce systemic effects on the central nervous system.

The chronic toxicity of a chemical in humans can be evaluated by three types of studies: 1) case reports, including histories of exposure to the toxic agent under investigation; 2) descriptive epidemiological studies, in which the incidence of various diseases in the population is found to be related by time or space to exposure to the agent; 3) analytical epidemiological studies, in which individual exposure to the agent is found to be specifically associated with an increased risk for certain diseases. Since people are usually exposed to multiple chemical and other risks simultaneously, including genetic factors, it is frequently difficult to unequivocally establish the degree of hazard posed by any single chemical. This is particularly true when dealing with the effects of low-level, long-term exposure.

CHAPTER 4

BIOTRANSFORMATION

4.1 THE ROLE AND COMPONENTS OF BIOTRANSFORMATION

4.1.1 GENERAL COMMENTS

Drugs or outright toxic substances, once inside the body, induce a defensive reaction from the body, however, chemical pollutants often do not induce such a response. Thus, the body's defense, as well as the effects of the pollutant are strongly related to metabolic and detoxification processes (biotransformation). It follows, therefore, that biotransformation of chemical pollutants is part of the complex overall process of nutrient assimilation and removal of toxic compounds. The metabolism of drugs and chemical compounds is one of the vital functions of the body's defense against invasion by foreign substances that may enter the body separate from or together with nutrients.

The earliest cells were forced to evolve enzymatic mechanisms to screen out or protect from toxic or non-assimilable compounds. The separation of nutrients from toxic substances involves the sensory system and barriers to penetration of the body. Due to their ability to sense and avoid or detoxify toxic materials, organisms were able to survive and adapt to an amazing variety of harsh environments. As mentioned in the first chapter, the same enzymatic systems that allowed our ancestors of 100,000 years ago to survive are present in our own bodies. This process, as defined by evolution, is the driving force of life; the ability to adapt to an increasingly complex environment.

The metabolism of foreign (xenobiotic) compounds generally involves the process of detoxification. Detoxification by conjugation with reduced glutathione or glucuronic acid is a major step in the metabolism of foreign compounds. Elimination of the products of metabolism and detoxified compounds is through excretion into the bile followed by elimination in the feces, or by transport to the kidneys followed by elimination in the urine. Elimination of some compounds can also occur through the skin (sweat) and lungs (expired air).

In general, the detoxification of foreign compounds includes two phases. The first involves hydroxylation or oxidoreduction. In this phase, additional OH, NH_2, or COOH groups are attached to the foreign compound. This phase can be considered as preparatory, increasing the polar character of the compound and making it more susceptible to being metabolized in the next phase. The products of the first phase may be pharmacologically active or toxic, while the products of the second phase are, as a rule, inactive [65, 286].

The second phase involves conjugation of the first phase products with natural compounds, such as glucuronic acid, and excretion of the conjugated compounds in the feces, urine, sweat, and expired air. The conjugation of chemical pollutants masks the functional groups introduced in the first phase, inactivating the molecule, decreasing its biologic activity. The lipid-soluble nature of certain chemical pollutants favors their passage of the cellular membrane, while water soluble nature favors the compound's being excreted. Substances with a high lipid/water partition coefficient readily pass through biological membranes, but will also diffuse back from the urine into the plasma. These substances have a low renal clearance and have a marked persistence in the organism.

The group of enzymes involved in the first phase (monooxygenases) evolved early in the history of life and continues to be important in the metabolism of toxic materials. Several authors [63, 65, 147, 287] suggest this group of enzymes appeared through evolution as a consequence of the metabolism of steroid hormones, but also provided a defensive weapon against foreign compounds present in the environment.

Table 4.1
Characteristics of metabolizing enzyme systems

Characteristic	Type of compound:	
	Nutrients	Toxins
Subcellular location	Mitochondria	Endoplasmic reticulum
Substrates	Water and lipid soluble	Lipid soluble
Enzyme variation	High	Low
Enzyme quantity	High	Low
Enzyme specificity	High	Low
Metabolic regulation	Allosteric	Genetic (Induction)

In table 4.1 the main differences between enzymes involved in the metabolism of nutrients and toxic materials are presented. The main purpose of monooxygenase is the transformation of any lipid soluble compound to a more polar derivative, which can be more easily eliminated. Without monooxygenase, there would be an accumulation of lipid soluble compounds in the tissues with toxic or other adverse biological effects. This role may explain the low specificity of these enzymes, as well as their inducibility and involvement in the metabolism of some natural compounds such as steroid hormones. While both nutrients and toxic compounds are metabolized, this transformation does not necessarily occur in the same organs or tissues. The liver does metabolize both classes of compounds, but not always using the same enzyme systems.

As shown in figure 4.1, chemical pollutants, drugs, and foreign organic compounds are metabolized using a variety of enzymatic systems, depending on the chemical characteristics of the compound. For example, an organic chemical pollutant (CP) can be metabolized to three different products (C, C', C") with different biological activities, depending on the nature of the active enzymatic system. As is the case for drugs, metabolism may involve more than one of these paths, explaining the presence of secondary products or effects [49, 287].

Most of the reactions occurring in the first phase of biotransformation occur in the liver. Actually, all tissues are active in the process of biotransformation, but the liver is the most active of the organs. In the liver, all subcellular fractions are involved in biotransformation reactions.

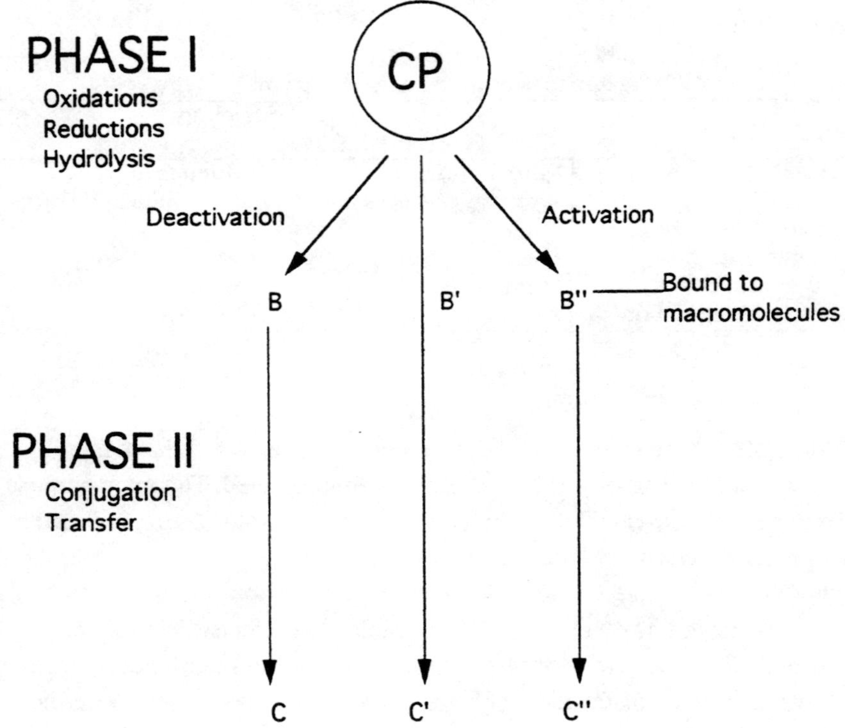

Figure 4.1. Possible detoxification pathways of a xenobiotic.

4.1.2 MONOXYGENASE

As previously mentioned the metabolism of foreign compounds developed in the course of evolution, probably with the transition of life from the aquatic to the terrestrial environment. Marine organism, as a rule, does not possess such metabolizing systems, but excrete lipid soluble compounds directly into the water.

The endoplasmic reticulum (ER) of liver and other cells appears microscopically as a tubular network in the cytoplasm. Two forms may be detected. Rough endoplasmic reticulum is covered with ribosomes, and smooth endoplastic reticulum, which is free of ribosomes but is rich in enzymatic activity. The synthesis of enzymes and other proteins occurs in the rough ER, which can loose its ribosomes and become smooth ER. During homogenization of tissues, cells rupture and the ER becomes broken down into small particles known as microsomes. The microsomal fraction can be obtained by differential centrifugation at 105,000 g for 60 minutes, after the other cell constituents have been removed. Under these conditions the microsomes sediment, leaving the cytosolic supernatant, also called the soluble fraction. Thus, microsomes are, in fact, experimental artifacts obtained *in vitro* by the destruction of the cellular structure.

The classic work of Brodie drew attention to the biochemical transformation of many foreign compounds of varied structure that takes place in liver microsomes. The requirement for molecular oxygen and NADPH (as the main coenzyme) suggested that the later formed a complex with oxygen, transferring it to the organic compound. An additional factor, later identified as cytochrome P_{450}, forms part of this complex. The name of this cytochrome is derived from its ability to combine in the reduced state with carbon monoxide, giving a characteristic absorbance peak at 450 nm.

Cytochrome P_{450} (a monooxygenase) is the site of oxygen activation. It functions as a monooxygenase in the oxidation of a large variety of substances such as steroids, cholesterol, fatty acids, drugs, polycyclic hydrocarbons, aromatic compounds, etc. (table 4.2). Its activity converts lipid soluble substances into hydrophilic derivatives by hydroxylation.

There are several possible hydroxylations possible for any given organic compound. Evolution settled on the use of cytochrome P_{450}. Several workers have studied the evolution of monooxygenase [49, 103, 78]. Monooxygenase systems possess few components, but are able to metabolize a wide range of substrates through a variety of reactions (oxidation, reduction, hydrolysis, dealkylation, and epoxidation).

Cytochrome P_{450} isolated from different sources (mammals, insects, bacteria) has a molecular weight that ranges from 47,000 - 60,000 daltons. When active, cytochrome P_{450} forms a complex that includes the substrate and O_2. To accomplish this, the Fe^{2+} contained in the cytochrome molecule requires two electrons. These are provided by the electron transport chain in the form of NAHPH. The transfer of electrons requires a reductase (NADPH cytochrome P_{450} reductase).

Table 4.2
Types of reactions catalyzed by biotransformation enzymes

Substrate	Example
Tertiary amines $X\text{-}N\text{-}R_2 + O_2 \rightarrow X\text{-}NO\text{-}R_2$	Nicotine, benzodiaepine
Secondary amines $X\text{-}N\text{-}R + O_2 \rightarrow X\text{-}NOH\text{-}R$	Methamphetamine
Arylamines $R\text{-}C_6H_4\text{-}NH + O_2 \rightarrow R\text{-}C_6H_4\text{-}NOH$	2-Naphthylamine
Thiocarbamides $X\text{-}N\text{-}CS\text{-}N\text{-}R + O_2 \rightarrow X\text{-}N\text{-}CSOH\text{-}N\text{-}R$	Thiourea
Thiols $2X\text{-}SH + O_2 \rightarrow X\text{-}S\text{-}S\text{-}X$	Mercaptoethanol
Sulfides $X\text{-}S\text{-}R + O_2 \rightarrow X\text{-}SO\text{-}R$	Dimethylsulfoxide
Thioamides $X\text{-}CS\text{-}N\text{-}R + O_2 \rightarrow X\text{-}C(S=O)\text{-}N\text{-}R$	Thioacetamide

When certain substances are added to a microsomal suspension in the presence of oxygen, but without NADPH, a characteristic type I spectrum is obtained. Substances that produce a type I spectrum include hexobarbital, aminopyrine, and testosterone. A type I spectrum has a peak at 385 - 390 nm and a minimum at 420 nm.

Incubation with aniline or pyridine produces a type II spectrum, with a peak at 426 - 435 nm and a minimum at 395 nm. Incubation with certain carcinogenic compounds (i.e., methylcholanthrene) produces a special spectrum with a peak at 448 nm.

All monooxygenases exhibit genetic variation and isoforms of each component have been isolated. The concentration and activity of cytochrome P_{450} may vary considerably from one species to another, or in the same individual with age, hormonal influence, stress, etc.

The electron transport system found in the microsomes is of great importance for the first phase in the metabolism of drugs or chemical pollutants. As in the mitochondria, this system transfers electrons or reducing equivalents along a sequence of redox carriers. The microsomal system starts with NADPH and ends with cytochrome P_{450}, which is the terminal oxidase.

Oxygen is bound to the heme portion of the cytochrome, where it forms short lived, labile hydroperoxides with the substrate. Before the product is released, the hydroperoxide is reduced to the hydroxyl.

In addition to the above mentioned components, a complete monooxygenase system requires cytochrome b5, cytochrome b555, various reductases, phosphatidylcholine, a protein carrier containing iron, and some other enzymes such as glucose-6-phosphatase. Monooxygenase systems require a stable membrane to function. Cholesterol functions to stabilize membranes and represents approximately 6% of the lipids in the endoplasmic reticulum.

Monooxygenase is part of the structure of the smooth ER. In close proximity will be found the components of the second phase of biotransformation, including glucuronyl transferase, glutathione S-transferase, and epoxide hydrolase. Globular protein molecules, like enzymes, are inserted within or across the phospholipid bilayer, creating macromolecular clusters (Fig. 3.5).

The electron transport chain together with the redox cycle of the cytochrome create activated oxygen as part of the normal function of this system. The activation of oxygen, which is required to complete the first phase of biotransformation, involves the formation of reactive oxygen free radicals (superoxide, $O_2^{\bullet-}$, sometimes written $O_2^{\bullet}$ or $^{\bullet}O_2$). In the presence of polyunsaturated fatty acids, and iron containing proteins, conditions are favorable for lipid peroxidation. Of all the subcellular fractions, microsomes from any organ will readily produce lipid peroxides *in vitro*. It is this effect that makes the metabolism of foreign compounds such a significant potential source of membrane damage. In fact, lipid peroxidation is a common consequence of contamination with chemical pollutants.

4.1.3 ENZYME INDUCTION

If phenobarbital or polycyclic aromatic hydrocarbons (generally carcinogenic) are added to a medium containing liver homogenate, no increase in the activity of the microsomal enzymes is observed. An increase is only noted *in vivo* a few days after the injection of these substances. Today, more than 300 compounds that increase the activity of these enzymes are known. Almost all the enzymes involved in biotransformation are inducible, having very low normal concentrations in the endoplasmic reticulum. As mentioned, this fraction is also rich in ribosomes, where protein synthesis occurs. Since known inhibitors of protein synthesis blocked the increase in enzyme activity, it is established that *de novo* synthesis of microsomal enzymes is involved in the induction of these enzymes. Actinomycin D, which inhibits the synthesis of messenger RNA, likewise prevents enzyme induction.

Enzyme induction is a complex process of the organism's adaptation to the presence of lipid-soluble substances of both exogenous and endogenous (steroids) sources. Thus, induction is a defensive weapon used by the organism to provide, by protein synthesis, needed enzymes to detoxify or metabolize the inducing substance. Unlike other types of enzyme regulation, such as activation by metal ions or feedback regulation, which have rapid effects and function on a short time scale, induction requires exposure to the inducing substance for hours or days. Intitation of induction varies depending on the nature of the inducing substance, particularly its concentration and half-life in the tissues.

Induction of microsomal enzymes is accompanied by an increase in liver weight, proliferation of smooth ER, and by an increase in the phospholipid and cytochrome content. Pretreatment of an experimental animal with an inducer will increase the metabolic rate of a great number of drugs and other chemical compounds. This increase in metabolic rate is concomitant with the increase in activity of NADPH cytochrome c reductase, cytochrome P_{450}, and the content of heme and phospholipids. At the same time there are structural changes in the endoplasmic reticulum that occur in parallel with hypertrophy of the liver.

Substances capable of inducing this response are highly lipid soluble, concentrate in the liver, and have a long half-life in the body. It is obviously impossible to present a complete list of inducing substrates. The main groups of compounds are given in table 4.3.

Table 4.3
Main groups of enzyme inducers

Drugs	Chemical pollutants	Steroids
Barbiturates (phenobarbital)	Carcinogenic hydrocarbons (methylchloranthrene)	Testosterone and derivatives
Benzodiazepines (diazepam)	Organochlorinated insecticides (lindane, DDT)	Cortisone
Neuroleptics (chlorpromazine)	Herbicides (paraquat)	Progesterone
Analgesics (morphine)	Polychlorinated diphenyls	
Antiinflammatories (phenylbutazone)	Compounds in smoke (marijuana)	
Antihistamines	Alkaloids (rye ergot)	
Anticoagulants	Food additives	

The practical importance of enzyme induction is the fact that chronic exposure to a drug or chemical pollutant increases the amount of the enzymes involved in its metabolism. Also, because of the low specificity of these enzymes, there will be a marked increase in the metabolism of other substances. As modern medical treatment uses a wide range of drugs, the induction effect can cause unexpectedly rapid metabolism of prescribed drugs. Clinicians are well aware of the interactions between drugs such as phenobarbital, and coumarin anticoagulants, or anti-inflammatory drugs (phenylbutazone).

Table 4.4
Phenobarbital and methylcholanthrene induction of rat liver enzymes, after
Li [169]

Enzyme	Substrate	% of control	
		Phenobarbital	Methylcholanthrene
Epoxide hydrolase	Styrene 7,8-oxide	206	162
	Naphthalene	230	165
	Benzopyrene 4,5-oxide	265	188
UDP-glucuronic transferase	Bilirubin	208	100
	Chloramphenicol	364	65
	p-Nitrophenicol	130	305
	Hydroxybenzopyrine	170	620
Glutathione-S-transferase	Morphine	230	125
	Styrene Oxide	140	205
	Chloro-2,4-dinitrobenzene	206	168

Ethanol and carcinogenic polycyclic aromatic hydrocarbons are strong enzymatic inducers. The administration of one inducer often increases the metabolism of other substances with related structures. There are several groups of inducing substances, including 1) most drugs and polychlorinated insecticides, 2) polycyclic hydrocarbons (methylcholanthrene), 3) ethanol. Comparative data concerning the effect of inducers are presented in table 4.4. The rats in this experiment were treated with phenobarbital (75 mg/kg/day) or with methylcholanthrene (25 mg/kg/day) for four consecutive days. The increase in enzymatic activity was generally greater for exposure to phenobarbital.

Enzyme induction has been demonstrated in microorganisms, mammals, amphibia, reptiles, and insects [6, 82, 183]. This process clearly occurs predominately in the liver, but has also been shown in kidneys, lungs, intestine, and skin. The main result of enzyme induction is the accelerated metabolism of undesired or excess lipid soluble substances (such as bilirubin). However, this process is actually a double-edged sword. The benefit is to help man and other organisms to survive in a complex environment where they face continuous exposure to compounds with varying toxicity. The damaging side relates to the nature of the products of this biotransformation. Most aromatic compounds are metabolized through several steps resulting in products with variable biological activity or toxicity. Carcinogensis is the most extreme potential negative effect of biotranformation. Another negative effect is the induction of antibiotic resistance in bacteria. The same effect has been discovered in solid tumors in humans, which become able to resist cytostatic compounds or high oxygen partial pressures by the induction of superoxide dismutase. Through this induction tumors are able to resist the increased exposure to free radicals resulting from the exposure to oxygen or anthracyclin antibiotics (cytostatics).

The discovery of enzymatic induction demonstrated the existence of adaptation at the molecular level. This had a great impact on biochemistry and pharmacodynamics in understanding the mechanisms of survival. The introduction of antibiotics that function as enzyme inhibitors provided another milestone. Many chemical pollutants can be enzyme inhibitors and inducers. This raises concern about possible interactions between drugs and environmental pollutants.

4.2 FIRST PHASE REACTIONS

Metabolism of lipid soluble chemicals takes place in the liver. First phase reactions increase the polarity of the molecule by adding oxygen containing groups. In table 4.5 the principle reactions belonging to the first phase of biotransformation are described.

During the experimental determination of the metabolic rate of chemicals by liver microsomes, a number of systems are employed including *o*-demethylation of 1-nitroanisole in p-nitrophenol, and the hydroxylation of aniline in p-aminophenol. For monitoring dealkylation reactions, formaldehyde is commonly used [20, 147].

The complexity of products coming from first phase reactions is mainly due to the simultaneous metabolism of the xenobiotic substance along multiple pathways. This produces multiple derivatives with variable biological properties. For example, the biotransformation of chlorpromazine produces 20 derivatives having differing properties [287]. The same type of result can be demonstrated using a wide range of chemical pollutants such as polycyclic aromatic hydrocarbons, aflatoxin, insecticides, etc.

Table 4.5
Reactions catalyzed by biotransformation enzymes

Type	Reactions
Oxidations	
Dealkylations	$R\text{-}OCH_3 \rightarrow R\text{-}OCH_2OH \rightarrow HCHO + R\text{-}OH$ (phenols) $R\text{-}NHCH_3 \rightarrow R\text{-}NHCH_2OH \rightarrow HCHO + RNH_2$ (amines) $R\text{-}SCH_3 \rightarrow R\text{-}SCH_2OH \rightarrow HCHO + R\text{-}SH$ (thiols)
Hydroxylations	aliphatic: $R\text{-}(CH_2)_n\text{-}CH_3 \rightarrow R\text{-}(CH_2)_n\text{-}CH_2OH$ aromatic: $C_6H_5R \rightarrow C_6H_5OH$ or $C_6H_4(OH)_2$ (hydrocarbons)
Oxidations	$R_3N \rightarrow R_3\text{-}N\text{=}O$ (aniline) $R\text{-}S\text{-}R' \rightarrow R\text{-}(S\text{=}O)\text{-}R'$ (benzodiazepines)
Oxidative deaminations	$R\text{-}CH\text{-}NH_2 \rightarrow R\text{-}C\text{=}O + NH_3$
Desulfurations	$RP(X)\text{=}S \rightarrow RP(X)\text{=}O$ (thiophenol)
Epoxidations	$R\text{-}CH\text{=}CH\text{-}R \rightarrow R\text{=}CH(O)CH\text{-}R$ (insecticides)
Reductions	
Nitro to amine	$R\text{-}NO_2 \rightarrow R\text{-}NH_2$ (aniline)
Ketone to secondary alcohol	$R\text{-}C(\text{=}O)\text{-}R' \rightarrow R\text{-}CHOH\text{-}R'$
Aldehyde to primary alcohol	$R\text{-}CHO \rightarrow R\text{-}CH_2OH$ (acetaldehyde)

4.2.1 AROMATIC HYDROXYLATION

Aromatic hydroxylation is induced by a wide range of chemical pollutants and drugs including phenols, nitro and azo derivatives, and catechols. This reaction produces very unstable arene oxides as reactive intermediates (fig. 4.2 and 4.4). This class of compounds also includes benzene oxide, and

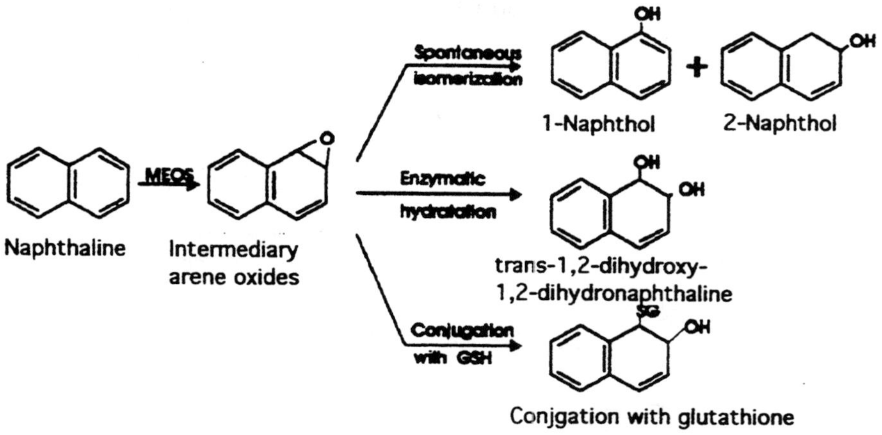

Figure 4.2. Liver metabolism of naphtalin and the formation of arene oxides.

epoxides, which undergo structural rearrangement to form phenols or dihydrodiols. This reaction is catalyzed by monooxygenases. Aromatic hydroxylation has been extensively studied. For aromatic rings larger than benzene, and especially for polycyclic aromatic hydrocarbons, it is known that aromatic hydroxylation can result in the formation of carcinogenic compounds [115, 187].

Epoxide hydrolase (EC 4.2.1.6.3) catalyzes the hydrolytic transformation of epoxides, which are very reactive mutagenic and/or carcinogenic substances, into their respective diols. This enzyme has a low substrate specificity. Epoxides are formed during the metabolism of some polycyclic aromatic hydrocarbons, especially those having carcinogenic or procarcinogenic properties.

The third reactive pathway of epoxides is conjugation with reduced glutathione (GSH). This reaction may occur nonenzymatically, but more frequently is catalyzed by GSH-S-transferase (EC 2.5.1.18), found in the cytosol. Aromatic hydroxylation is a complex process. It can be observed in all organisms from the bacteria upward.

There are other problems aromatic chemical pollutants can present to the body. A large number of substances used in industry, such as derivatives of aniline, nitrobenzene, b-naphthylamine, *o*-toluidine, benzidine, 4-aminostilbene, and primaquine, are capable of producing methemoglobinemia of varying severity. Unlike other compounds that cause the production of methemoglobin both *in vitro* and *in vivo* (like nitrates and hydroxylamine), these compounds are only active *in vivo* following biotransformation. It is the metabolism of these compounds that produces the true methemoglobinizing agents. The action of monooxygenase on these compounds produces *p*-aminophenol, *p*-nitrosophenol, and phenyl hydroxylamine. The formation of methemoglobin is dangerous, especially for people with a congenital deficiency in glucose-6-phosphate dehydrogenase, which controls the regeneration of NADPH and GSH.

The previously mentioned aromatic compounds are also involved in the production of some allergic effects, which are more frequently seen in industrialized areas. The mechanism of action involves the formation of active intermediates with haptene properties, due to covalent binding to plasma or tissue proteins. Such allergies, involving skin eruptions, have low specificity and cause hypersensitivity to other chemical pollutants [44, 176, 187].

4.2.2 REACTIONS IN OTHER SUBCELLULAR FRACTION

In addition to the first phase reactions that occur in the endoplasmic reticulum, other metabolic reactions take place in mitochondria and in the cytoplasm of cells.

OXIDATIVE DEAMINATIONS are catalyzed by mono- and diamine oxidases. The reaction products are aryl- or alkylaldehydes that are further oxidized to their corresponding acids by other enzymes. A typical reaction is:

$$R\text{-}CH_2\text{-}NH_2 \rightarrow R\text{-}CH=NH \rightarrow R\text{-}CHO + NH_3$$

Monoamine oxidase is particularly active in tissues where catecholamines are formed and stored (catecholamines are subsequently deaminated into *o*-methyl derivatives). Diamine oxidase acts on a large variety of substrates including histamine, polyamines (cadaverine), and benzylamine.

OXIDATIONS can occur in several places within the cell. The cytoplasm is rich in enzymes that catalyze the oxidation of aromatic and aliphatic aldehydes. These β-oxidations produce carboxylic acids.

A special role is held by ceruloplasmin, the single oxidase found in the plasma. This protein has eight atoms of copper and a low substrate specificity. It will act on Fe^{2+}, phenols, catecholamines, aromatic amines, chlorpromazine, etc. Ceruloplasmin is present in the plasma in a high concentration and is probably part of the detoxification process [83,211].

REDUCTIONS are catalyzed by reductases, which act on the following substances:

- monocyclic unsaturated terpens (β-phelandren)
- disulfides (cystamine)
- sulfoxides (dimethylsulfoxide)
- drugs (chloramphenicol, prontoside)

Hydrolysis of esters and amides involves the catalytic action of hydrolases from the plasma and tissues. Plasma esterases exhibit considerable genetic variation allowing them to act on a variety of drugs (aspirin, local anesthetics, procaine, cyclophosphamide, insecticides, etc.). Of the more than 50 known esterases, most are found in the intestine and are produced by the bacterial flora. b-glucuronidase uses the products of the second phase of biotransformation as substrates, which form glucosiduronic esters. Drugs, such as morphine and chloramphenicol or polycyclic aromatic hydrocarbons, are excreted as esters in the bile. Hydrolysis occurs in the intestine and the products are absorbed forming an enterohepatic cycle. Among the reactions that can occur are:

- splitting of aromatic rings as occurs with benzene
- dehalogenization as is observed with insecticides (DDT)
- acetylation of aromatic amines and sulfonamides in positions 1 and 4

Acetylation of metabolites occurs in the liver. Deacetylation of the compounds produced in the liver may occur in the kidneys.

4.3 SECOND PHASE REACTIONS

Through conjugation reactions, chemical pollutants and other xenobiotics or their metabolites enter the second phase of metabolism, increasing their hydrophilic nature, allowing for their excretion. Conjugation involves the combination of the metabolites of chemical pollutants with endogenous substrates, such as glucuronic acid, GSH, sulfates, or glycine. These conjugation reactions typically occur at the site of another functional group such as OH, NH_2, COOH, epoxide, halogen, etc. Conjugation is an enzymatic process and the enzymes are specific to certain types of reactions.

To accomplish the reaction, the substrate (either the conjugating agent or the metabolite) is activated by coupling with ATP. A specific transferase catalyzes the formation of the conjugated product. The first pathway, in which the donors are glucuronic acid and active sulfate, is:

Endogenous substrate $\xrightarrow{ATP}$ Activated substrate $\xrightarrow{Transferase}$ Conjugated product

In the second pathway the substrates are glycine and glutamate:

Exogenous compound $\xrightarrow{ATP}$ Activated substrate $\xrightarrow{Transferase}$ Conjugated product

The enzymes involved in the second phase of biotransformation have species specificity. In the lower life forms, conjugations are mostly made with carbohydrates or amino acids (glycine, ornithine, arginine). In man and the primates, conjugation with glucuronic acid and glutathione with conversion to a mercapturic acid prevails.

4.3.1 GLUCURONIDE FORMATION

Conjugation with glucuronic acid is the most widespread of the second phase pathways. It occurs in all mammals and vertebrates except fish. The formation of glucuronide takes place in two stages involving first, the biosynthesis of the glucuronyl donor: UDP glucuronic acid (UDPGA). This is followed by the transfer of the glucuronyl group from UDPGA to the metabolite being conjugated.

Conjugation reactions occur mostly in the liver, and to a lesser extent in the kidneys, gastrointestinal tract, and skin. The enzymes synthesizing UDPGA are found in the cytosol and glucuronyl transferase is found mostly in the endoplasmic reticulum.

This system acts in the microsomes on a variety of substrates including bilirubin, estrogen testosterone, thyroxine, phenolics, alcohols, etc. The product may be O-glucuronides (from aniline, sulfonamide, etc.) or S-glucuronides (from thiophenol, Antabuse, etc.). Conjugated glucuronides have a β configuration and are hydrolyzed by β-glucuronidase producing free glucuronic acid and aglycons. Glucuronides excreted through the bile may be hydrolyzed by intestinal glucuronidase and be reabsorbed. Thus, an enterohepatic cycle is present, as is the case for bilirubin.

The greatest activity of glucuronyl transferase is found in the liver, kidneys, intestine, and skin. As for other enzymes involved in detoxification, glucuronyl transferase is present as isozymes having differing affinities for different chemicals. Phenols and steroids form ether glucuronides, while bilirubin and anthranilic acid form ester glucuronides.

Glucuronyl transferase is strongly bound to the endoplasmic reticulum, keeping it in close proximity to cytochrome P_{450}.

As a rule, glucuronides are less reactive than the respective metabolites formed in the first phase of biotransformation. But there are exceptions. Aryl glucuronides produced from chemical pollutants seem to be involved in the etiology of bladder cancer [66, 76].

4.3.2 SULFOCONJUGATION (SULFATION)

A wide variety of compounds, including aliphatic alcohols, phenols, sulfamates, and steroids are eliminated by sulfation. The transfer of a sulfur group from 3'-phosphoadenosine-5'-phosphosulfate (PAPS) to a phenol, alcohol, or aniline uses the following enzymatic systems:

$$\text{ATP} + \text{sulfate} \xrightarrow{\ \textit{sulfurylase}\ } \text{Adenosine-5-phosphosulfate (APS)} + \text{Pyrophosphate}$$

$$\text{APS + ATP} \xrightarrow{\textit{APSkinase}} \text{PAPS}$$

$$\text{Phenol + PAPS} \xrightarrow{\textit{sulfotransferase}} \text{Phenyl-O-sulfate + PAP}$$

While sulfation is localized in the cytosol, it proceeds in parallel with glucuronic conjugation. Both pathways compete for the same substrates such as phenol, aromatic amines, and steroids. The actual competition is between glucuronyltransferase and sulfotransferase (EC 2.8.2.1). An equilibrium between the activities of these two pathways becomes established. For low levels of substrates, sulfation predominates. As the level of substrate increases, the rate of glucuronoconjugation also increases. Usually, sulfate esters are quite acidic and of low toxicity. The exception is the ester of N-hydroxy-2-acetylaminofluorene, which is reactive and strongly carcinogenic. It is important to know the ratio between conjugation and hydrolysis of sulfur esters by sulfatases. The presence of hydrolytic enzymes in the intestinal tract favors the formation of enterohepatic cycles. This results in the reentry of the toxic substance into the body [91].

4.3.3 CONJUGATION WITH GLUTATHIONE

In recent years, many studies have focused on the essential role of the tripeptide, glutathione (GSH), in life processes [65, 287]. GSH is also of great importance in detoxification processes. Its ability to conjugate to a great number of reactive substances suggests it may be the most important defense mechanism at the cellular level [40, 181, 182].

GSH functions in the detoxification process as a coenzyme for glutathione-S-transferase (EC 2.5.1.18). This is actually a group of enzymes that constitute 10% of the total protein extractable from rat liver. In addition to the conjugation of GSH to reactive compounds, this group of enzymes appears to play a role in other detoxification processes. One is the binding of a large number of xenobiotics as well as bilirubin, the toxic product of heme degradation, until they are converted to less toxic compounds. Another function is being a scavenger peptide. Many reactive intermediates are produced during biotransformation of chemical pollutants. GSH-S-transferase binds covalently to these compounds, preventing them from damaging the cell.

Hepatic GSH transferase is inducible by some compounds such as phenobarbital, 3-methylcholanthrene, hexachlorobiphenyl, tetrachlorodibenzo-*p*-dioxin, etc. Enzymatic activities appear to be age dependent and are subject to variations with hormonal changes. GSH transferase activity is usually measured with an aryl substrate (1-chloro-2, 4-dinitrobenzene) or an epoxide {1,2-epoxy-3- (p-nitrophenoxy)-propane}. Metabolic inducers (phenobarbital, DDT) or fasting lead to marked increases in transferase activities. Partial hepatectomy, acute and chronic alcohol pretreatment, and thyroid dysfunction result in a pronounced increase in the epoxide transferase activity, while the aryl transferase activity remains unchanged. Experimental blockage of the bile duct and liver preinjury followed by administration of carbon tetrachloride decreased the GSH transferase activity for the epoxy substrate, but not the aryl substrate [40, 182].

GSH transferases can be found in all subcellular fractions, but the cytosol is the richest with seven isoenzymes. GSH transferases have molecular weights of around 50,000 and are composed of two apparently identical subunits. However, the properties of the isoenzymes very greatly in regard to optimum pH and affinity for their substrates.

A partial list of substrates for GSH transferases is given in table 4.6, which illustrates the great structural variety of substrates for this enzyme. This property is due to the affinity of GSH for electophilic sites in chemical pollutants. While the substrate specificity of GSH transferase for xenobiotics is low, the specificity for GSH is very high. No other thiol can replace GSH as a cofactor for this enzyme [5, 103, 182]. The low specificity for xenobiotics can create problems in certain circumstances. For example, rat liver GSH transferase β is able to act both on nitroglycerin (1):

$$CH_2(ONO_2)CH(ONO_2)CH_2(ONO_2) + 2GSH \rightarrow$$
$$CH_2(ONO_2)CH(OH)CH_2(ONO_2) + GSSG + HNO_2 \qquad (1)$$

and on ethylthiocyanate (2):

$$CH_3CH_2SCN + GSH \rightarrow CH_3CH_2SSG + HCN \qquad (2)$$

Table 4.6
Substrates of GSH transferases grouped by enzymatic activity

Aryl transferase	Alkyl trasferase
-Mono and bichloro nitrobenzenes -4-Nitroquinoline (carcinogen) -Atrazine (herbacide) -Sulfobromophthalein -Naphthalene	-Organophosphates -Halogenated hydrocarbons -Ethylmethane sulfonate
Epoxide transferase -Polycyclic hydrocarbons -2,3-Epoxipropylphenyl -1,2-Epoxipropane -Styrene oxide	Alkyl transferase -Unsaturated carbonyls -Acroleine -Acrylic acid esters -2,3-Dimethyl phenoxyacetate -Methyl vinyl sulfonate -Diethylmaleate
Arakyltransferase -Benzyl chloride	Estrogen transferase -17-β-Estradiol

The activity of GSH transferase produces mixed disulfides. Keer and Jakoby [140] have proposed a mechanism to explain the reaction on nitrate esters such as nitroglycerin (reaction 1 above). They propose a reactive intermediate, sulfanyl nitrite (GS-NO$_2$) is formed, which reacts nonenzymatically with a second molecule of GSH releasing nitrous acid (HNO$_2$) and oxidized glutathione (GSSG).

$$GSH + RNO_2 \xrightarrow{\textit{transferase}} GS\text{-}NO_2 + ROH \qquad (3)$$

$$GS\text{-}NO_2 + GSH \xrightarrow{\textit{nonenzymatic}} GSSG + HNO_2 \qquad (4)$$

Reaction 4 looks like an interchange reaction between a disulfide and a thiol with formation of a mixed disulfide (GSSR)

$$GSSG + RSH \rightarrow GSSR + GSH \qquad (5)$$

Such interchange reactions of GSH are common and may proceed nonenzymatically or be catalyzed by transhydrogenases. Such a mechanism may explain the inhibitory action of thiols and the great specificity for GSH. GSH transferases are inhibited by quinones or organic compounds containing Sn, Ge, or Cd (Ki about 3 µM).

The high activity of GSH transferase from rat liver may explain the resistance of this animal to chemical pollutants and various poisons. We should remember that the activity of this enzyme in man is about 5 fold lower [181, 190].

Acrolein is an aldehyde formed in low amounts during the cooking of meat. This compound has a high cytotoxicity due to the low dissociation of its mixed disulfide with GSH. This also leads to its remaining in the body for 3 - 5 days. As an indication of the toxicity of this compound, the limit on atmospheric discharge and presence in tap water is 2×10^{-9} M (0.013 - 0.043 ppm) [295]. Acrolein is also a very potent oxidizing agent, reacting nonenzymatically with free sulfhydryl groups (GSH, cysteine).

An organism's attempt to rid itself of mixed disulfides can lead to the formation of toxic materials due to acetylation by N-acetyltransferase. Thus it is useful that the metabolism of compounds such as paracetamol, benzoyl chloride, or polycyclic aromatic hydrocarbons involves several enzymes leading to the formation of differing products in varying amounts.

Exogenous compounds containing aromatic rings, especially halogenated nitrobenzenes, are excreted through the urine as acetylated cysteine derivatives. The formation of mercapturic acids (figure 4.3) takes place during enterohepatic cycling of bilinary compounds and in the kidney. Nitroquinoline (an alkaloid of the betel nut, used as a cathartic and taeniacide) participate in the formation of mercapturic acids [159, 182].

As will be in a later chapter, a dramatic drop in GSH levels follows exposure to many xenobiotics. Therefore, the maintenance of an adequate level of this compound is an essential condition for the survival of organisms in a polluted environment.

4.3.4 OTHER CONJUGATION SYSTEMS

The conjugation of aromatic chemical pollutants with amino acids, especially glycine, is fairly widespread, leading to the formation of hippuric acids. The conjugation of glycine and benzoic acid was discovered by Keller

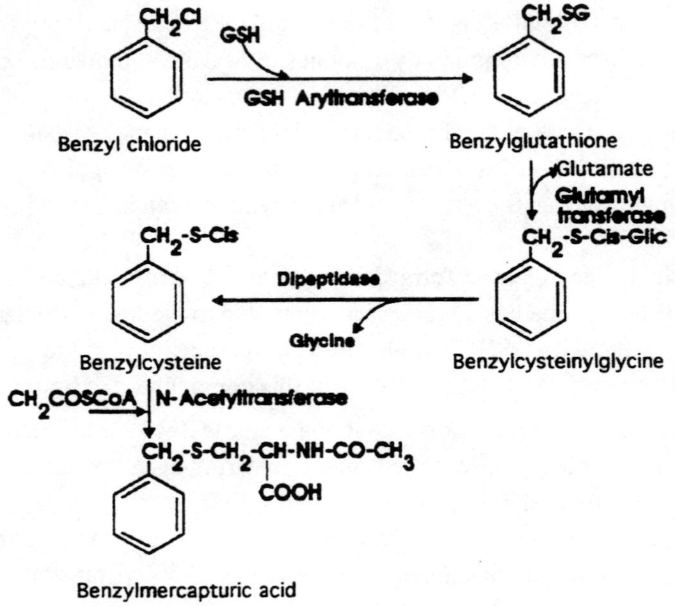

Figure 4.3.   Glutathione in biotransformation.

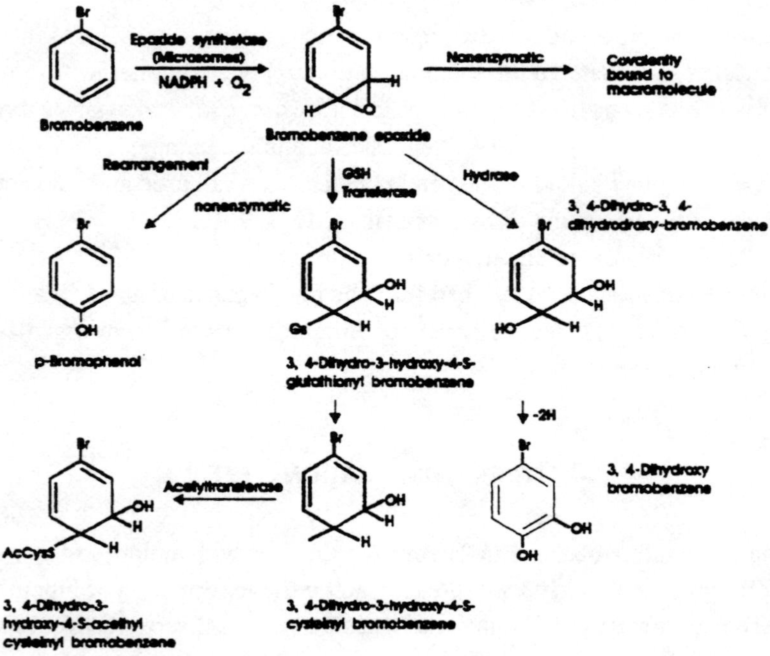

Figure 4.4.   Pathways for the metabolism of an aromatic compound.

in 1842. The production of hippuric acid was used as a liver function test and was initiated by the ingestion of 6 g of sodium benzoate. A urinary elimination of 4.0 - 4.5 g/4 hours was considered within the normal range.

The formation of hippuric acid requires coenzyme A and ATP as an energy source. In addition to benzoic acid, glycine may be consumed for conjugation to other aromatic compounds with a COOH group (benzenic acid, naphthalene, pyridine, thiophen, furan, etc.) or with unsaturated fatty acids (phenylacetic and cinamic acids). Conjugation using glutamine is limited to arylic acids (phenylacetic and indole-3-acetic acids).

N-acetyltransferase (EC 2.3.1.5) has a molecular weight of 37,000 and has two SH groups at the active site. The enzyme has a low specificity, acetylating a large range of compounds that contain NH_2 or -NH-NH_2 groups bound directly to an aromatic ring (aniline, hydrazines, sulfanylamides, p-aminobenzoic acid, etc.). The acetylated products possess the general structure of aromatic ring-NH-CO-CH_3, and have low aqueous solubility. The activity of N-acetyltrasferase varies between individuals, with both fast and slow acetylators being described [49]. The rate of acetylation in an individual may affect the rate of clearance of certain drugs.

Both natural products, like histamine, choline, noradrenaline, and nicotinic acid, and xenobiotics, like alkaloids, and p-dimethylaminoazobenzene, are methylated by the body. Methylation involves transfer of the methyl group from S-adenosylmethionine to amines, phenols, and thiols, forming N-, O-, and S-methyl conjugates. S-adenosylmethionine is biosynthesized from methionine and ATP. The methyl group is then transferred to acceptors, such as catecholamines and psychogenic amines (bufotenine) by methyltransferases.

Among chemical pollutants, methylation is a means of eliminating aromatic compounds (quinoleine, pyridine, etc.) and some metallic ions (Hg, Pb, Cr, Se, As, Au, Sn, and Tl). For these reactions three cofactors are involved: S-adenosylmethionine, N-methyltetrahydropholate derivatives, and vitamin B_{12}. Methylation is very important in the elimination of contaminating metals, especially mercury. The process will operate with all metals having a redox potential in the range of +0.85 to +1.5 volts. For arsenic and thiols, methylation involves the formation of free radical intermediates [167].

Multiple metabolic pathways for chemical pollutants are not limited to conjugation, but also occur with other systems such as hydroxylation. In fact, phases I and II of detoxification are not strictly delimited. Generally, xenobiotic compounds are metabolized by multiple pathways, although one will predominate. As is the case for many natural compounds, necessary cofactors for one pathway may be lacking or in low supply. The presence of multiple pathways allows the reaction to proceed, and a normally minor pathway may temporarily become a major pathway. It is difficult to follow this type of adaptation in phase I of biotransformations, but in conjugation reactions, chromatographic analysis of the excreted metabolites often allows qualitative and quantitative measurement of the activity of various pathways [181].

For the most part, all conjugation pathways are available for the detoxification of xenobiotics. This has been most thoroughly documented in studying drug metabolism. Thus, p-aminosalicylic acid (PAS), used in the treatment of tuberculosis, is excreted without modification, as acetamidosalicylate, as p-acetamidoglucuronide, as p-aminosalicyluric acid, and simultaneously as p-aminosalicylglutamic acid.

4.4 MARKER ENZYMES FOR TOXICITY

The process of biotransformation should be regarded as a dynamic system of biochemical reactions by which the organism tries to detoxify xenobiotic compounds, or to eliminate natural waste products. In addition to examining its remarkable efficiency, one should also consider the negative aspects of biotransformations. First, the capacity for biotransformation is limited and can be overwhelmed by a high exposure to chemical pollutants. Second, as described in figures 4.1 to 4.3, most xenobiotic compounds or their metabolites are involved in several metabolic pathways. Very often, some of these metabolites reenter the pathways of biotransformation. Consequently, the measurement of these metabolites can give varied results, making interpretation difficult [159, 145].

It is worthwhile to consider the consequences of some enzyme activities involved in biotransformation.

Epoxide hydrolase (EC 3.3.2.3) converts epoxides into their respective diols. Some of these diols may be reoxidized by cytochrome P_{450} into diol-epoxides. These are strong mutagenic and carcinogenic compounds because of their capacity to bind covalently with proteins and nucleic acids. This cytoplasmic enzyme has a molecular weight of 100,000 and is inducible by a variety of compounds such as the procarcinogenic polycyclic aromatic hydrocarbons (PAH) and some drugs (phenobarbital). Epoxide hydrolase catalyses the addition of a water molecule at the epoxide group. For PAH, the addition of water takes place in the K region of the molecule, where a high electron density is involved in covalent binding of the molecule with nucleic acids. Therefore, as proposed by Kizer [142], epoxide hydrolase may be a marker for chemically induced liver cancer.

During the metabolism of PAH, a modified cytochrome (cytochrome P_{448}) is induced by the product of aryl hydrocarbon hydroxylase (EC 1.14.1.1) (AHH). As is the case for other enzymes involved in biotransformation, AHH may be induced by natural and xenobiotic substances containing phenyl, indol, or histidine rings. These compounds include catecholamines and their precursors (tyrosine), serotonin and its precursor (tryptophan), testosterone, and prostaglandins. The most potent inducer is 2,3,7,8-tetrachlorodibenzo-p-dioxin, an intermediate in the production of the herbicide 2,4,5-T. As shown in figure 4.5, dioxin significantly induces AHH activity at a concentration as low as 10^{-10} M. This is also the lethal dose. In comparison, the carcinogenic compound, methylcholanthrene, induces AHH at a concentration of 10^{-8} M.

The work of Poland and Kende [223] outlined the action of d-aminolevulinic acid synthase (ALA synthase) (EC 2.3.1.37), which catalyses the synthesis of d-aminolevulinate. ALA synthase is the rate-limiting enzyme in the biosynthesis of the porphyrine ring, which is critical to the formation of hemoglobin, cytochromes, catalase, etc.

As described in figure 4.5B, dioxin induces ALA synthase at the same low concentrations that induce AHH. ALA synthase is also induced by barbiturates, alyl-isopropyl-acetamide, and metals such as selenium and cobalt. In lead toxicity the activity of liver ALA synthase is inhibited. The inducing effect of some xenobiotics towards ALA synthase may explain the presence of some types of porphyria as a consequence of exposure to them.

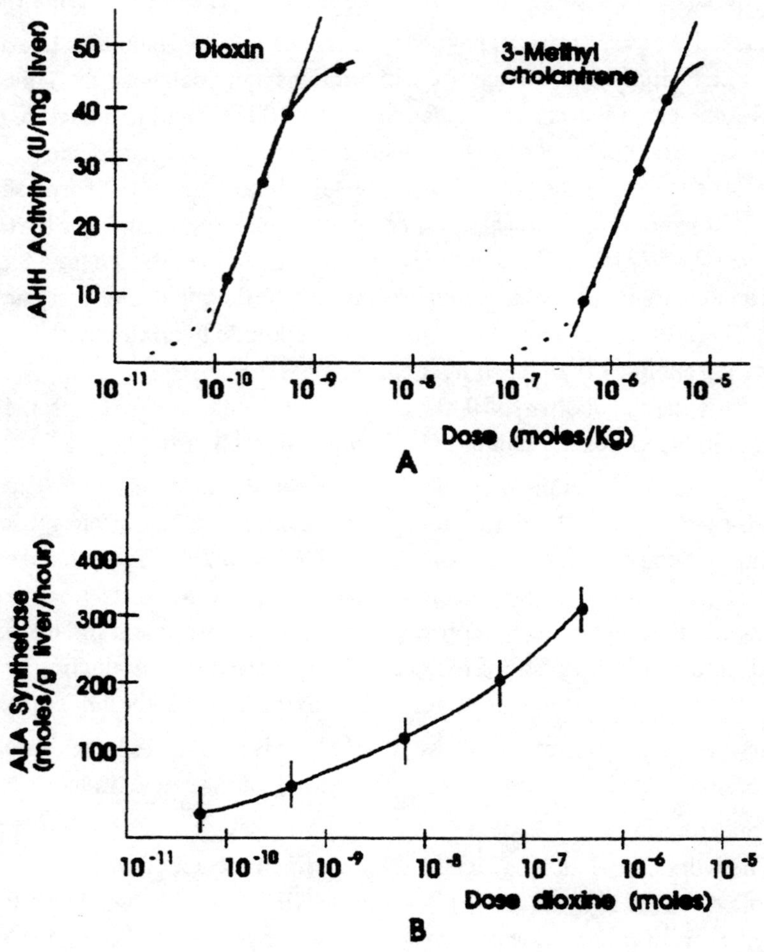

Figure 4.5. Induction of enzymes by dioxin. After Poland and Keade [223].

Certain effects of intoxication by xenobiotics are closely related to biosynthesis and degradation of hemoproteins. As previously stated, some chemical pollutants, such as nitrates and nitro derivatives, produce methemoglobinemia by a direct action on hemoglobin. This effect is particularly strong in people with a congenital deficiency in the enzyme glucose-6-phosphate dehydrogenase. Heavy metal ions, such as Pb, Cd, Co, and Se, affect both ALA synthase and heme oxygenase (EC 1.14.99.3). Heme oxygenase breaks down heme into biliverdin, CO, and Fe^{3+}. Heme oxygenase is also induced by the administration of hemoglobin, hydrazine derivatives, endotoxin, alcohol in high doses, or with erythrocyte lysis as occurs in metal poisoning [273]. Buttar [33] has shown that the induction of heme oxygenase parallels the destruction of cytochrome P_{450} by toxic compounds.

Continuous exposure to chemical pollutants increases the level of cytochrome P_{450} and other components of monooxygenase systems. An interesting study on the adaptation of wild birds exposed to pollutants is presented in table 4.7. For the first three species shown, there is a clear relationship between cytochrome levels and the amount of pesticide.

It is nearly impossible to prevent interactions between chemical pollutants and drugs. For example, patients treated with anticonvulsive drugs (diphenylhydantoin), that are enzyme inducers and also exposed to insecticides have less insecticide stored in adipose than people not taking the drug. The American Institute for Health and Environment has studied the effect of oral administration of drugs, insecticides, and carcinogens on rats. After a week of treatment, the level of cytochrome P450 and other components of monooxygenase systems were measured. The data presented by Bick and Fishbein [24] indicate:

- the inductive effect of 1 week of exposure was the same as that obtained with 14 weeks of exposure
- among the administered compounds, only DDT and phenobarbital were acting as true inducers
- no cumulative effect was seen, indicating all compounds were metabolized by the same pathway
- toxicity was characterized by an increase in liver weight, inhibition of some enzymatic components of the monooxygenase system, and a nonspecific increase in cytochrome P_{450}.

These results were only obtained when the animals were exposed to hepatotoxic compounds (carbon tetrachloride, thyoacetamide, and chloroform).

Table 4.7
Biotransformation enzyme and pesticide concentrations in the liver of wild
birds, after Yspa [302]

Species	Pesticide conc. (ppm)		Cytochrome P450 (nmoles/mg prot.)	Epoxidation of Aldrin
	Dieldrin	Polychlor		
Crow	1.2±0.8	0.77±0.5	0.44±0.21	2.43±0.3
Starling	0.91±0.3	0.61±0.5	0.51±0.20	1.98±0.2
Sparrow	1.11±0.9	0.47±0.3	0.27±0.08	2.45±0.3
Quail	0.37±0.2	0.34±0.2	0.18±0.07	1.02±0.6
Owl	0.39±0.1	0.41±0.3	0.34±0.13	1.00±0.3
Blackbird	0.42±0.27	0.54±0.4	0.36±0.15	0.92±0.3

These and similar studies [112, 145] show an adaptive biological response to continuous exposure to chemical pollutants in the environment. Such studies are not possible on humans, but evidence of similar adaptations and by-products can be detected in blood and urine [88, 124].

Transaminases and aminotransferases (EC 2.6.1.1 and 2.6.1.2) are the most used marker enzymes used in toxicological studies. Their low specificity is known. Siest [261] showed that for healthy people, the variation in transaminase activity due to age, sex, and weight may be over 50%. In comparison, a meal produces a variation of 2%, and anticonvulsive and anticonceptive drugs a variation of 36%. Elevated transaminases indicate cytolytic or necrotic effects of toxic exposure, liver disease, or heart failure.

Some studies on leucine aminopeptidase (EC 3.4.11.1), lactate dehydrogenase (EC 1.1.1.27), and alkaline phosphatase (EC 3.1.3.1) showed an elevation in levels 3 - 5 hours following experimental exposure to gentamicin or uranylactate [261, 268].

At the "Trends in Biological Dosimetry" congress held in Lerici (Italy) in 1990, a great deal of attention was given to the work of D. W. Britton, et al. and A. L. Burlingame, et al. concerning new types of analysis based on protein adducts to chemical pollutants. These methods use gas chromatography and mass spectroscopy to identify and characterize adducts of genotoxic carcinogens with hemoglobin or DNA. Genotoxic compounds bind to nucleophilic sites on cellular proteins and nucleic acids. C. S. Gionetti and J. Taylor of Argonne National Laboratory introduced a two-dimensional gel electrophoresis method to study the hundreds of individual proteins found in a single biological sample. This makes it potentially possible to detect changes in protein patterns caused by exposure to chemical pollutants that could serve as indicators of toxic effects. Such methods also require sophisticated image and data analysis software [169, 202, 281].

Human exposure to and absorption of pollutants can be measured either directly or indirectly. In the direct method, people carry a personal air pollution monitor and record their locations and activities over the course of the study. They also record water consumption and keep samples of the water from home, work, etc. Finally, they collect samples of their food for analysis. From this data, exposure, metabolic rates, and respiration can be calculated. In the indirect method, people keep an accurate, detailed diary of their locations and activities over the study period. This diary includes the amount of water drunk, and the amount and type of food eaten. For each location and activity an estimate is made of the air quality from values determined at similar locations. The exposure due to food and water is estimated from available data or by collecting and analyzing samples from that local area.

CHAPTER 5

ANTIOXIDANT PROTECTIVE SYSTEMS

5.1 OVERVIEW OF ANTIOXIDANTS

During evolution, the increased chemical complexity of the environment came with increased sources of oxygen activation. As described in an earlier chapter, oxygen activation results in the formation of harmful free radicals. This presented a challenge to organisms to adapt to the new situation by developing a variety of antioxidant systems. Consequently, all aerobic cells now possess several varied, efficient antioxidant systems. The characteristics of these systems are as follows.

a. They have both high and low specificity, having the ability to act at several critical points in the oxygen activation process. Some act enzymatically, others nonenzymatically. The natural antioxidants do not belong to a single chemical class of compounds. The lack of a common structure provides for a rapid, efficient and nonspecific action against oxygen activation.

b. They are metabolically connected, allowing their regeneration (e.g., the regeneration of reduced glutathione at the expense of ascorbate).

$$\text{Ascorbate} + \text{GSSG} \xrightarrow{2H+} \text{Dehydroascorbate} + 2\text{GSH}$$

Glutathione and its dependent enzymes are also closely connected to glycolysis to assure its continuous regeneration.

c. They have a high reaction rate. For example, SOD possesses a rate constant of 2×10^9 mol/sec, assuring a high efficiency. In cells with a high level of oxygen activation (erythrocytes, hepatocytes, etc.) there are higher concentrations of antioxidant enzymes.

d. The specificity of individual antioxidant enzymes is generally low, but the efficiency is increased by several enzymes functioning synergistically (i.e., SOD and catalase).

e. In cells, antioxidant systems are localized near sites of oxygen activation, especially the membrane, mitochondria, and endoplasmic reticulum. Each subcellular fraction contains antioxidants in differing combinations and amounts, depending on the nature of reactions occurring there.

f. In some cases, nonspecific antioxidants are stored against future need (carotenes or metallothioneins).

The major enzymatic and nonenzymatic antioxidants are shown in figure 5.1. This scheme shows the relation between the steps of oxygen activation, the activities of antioxidants, and some of the regenerative capacity of antioxidants. While enzymes with high reaction rates (SOD, catalase) tend to act independently, those involved in the decomposition of peroxides are metabolically connected to glycolysis to assure an adequate supply of NADPH or they regenerate each other (vitamins C and E or glutathione).

5.2 SUPEROXIDE DISMUTASE (SOD)

Although there are many natural substances and enzymes capable of reacting with the superoxide radical ($O_2^\bullet$), only SOD evolved to serve this role as an antioxidant. Because SOD can be found in nearly all aerobic organisms, it has been proposed that its function is the first line of defense against superoxide-mediated injury. The only known substrate for SOD is $O_2^\bullet$. Therefore, SOD is a specific catalyst that has evolved to remove $O_2^\bullet$ by converting it to H_2O_2:

$$2O_2^\bullet + 2H^+ \xrightarrow{\;SOD\;} H_2O_2 + O_2 \tag{1}$$

The H_2O_2 produced can be removed by catalase (EC 1.11.1.6) by the following reaction:

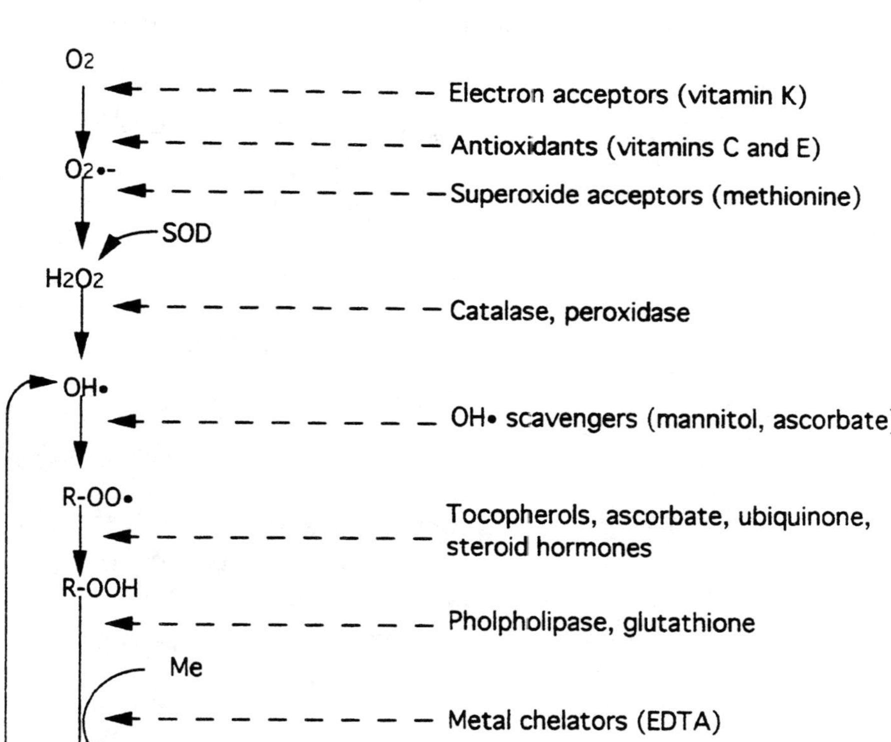

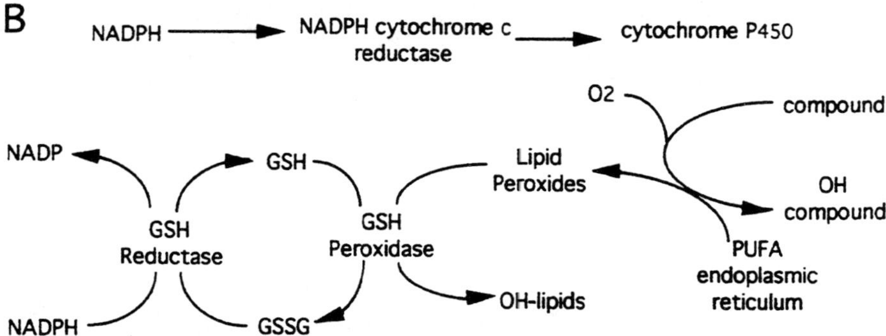

Figure. 5.1. The main natural antioxidant systems A: Action of antioxidants during oxygen activation. B: Relationship between natural antioxidant systems.

$$2H_2O_2 \xrightarrow{\text{Catalase}} 2H_2O + O_2 \qquad\qquad (2)$$

SOD (EC 1.15.1.1) possesses metallic ions in the active center. Cu-Zn SOD is found in the cytoplasm of eukaryotic cells. This enzyme was discovered by Fridovich and McCord, of Duke University, North Carolina, in 1969. This SOD is composed of 2 similar subunits, each containing a zinc and a copper ion. It has a molecular weight of 33,000.

A second SOD contains 2 - 4 manganese atoms per molecule. It has 4 subunits and a molecular weight of 80,000. This SOD is found in mitochondria and in most bacteria.

In 1982, Markland, of Uppsala, Sweden, discovered a third SOD in the extracellular space of pulmonary tissue. This SOD possesses 4 copper and 4 zinc atoms with a molecular weight of 135,000.

A large amount of literature has been written about this enzyme [33, 180]. Consequently, we will only briefly discuss the features of the enzyme.

CuZn SOD is the most common form of this enzyme. It is found at high concentrations in cells and tissues having an intense metabolic activity. An exception is activated leukocytes, which have high metabolic activity but low a SOD level. This may be compensated for by the very high level of catalase in these cells. Erythrocytes have a relatively high level of SOD, which is constant for all species of mammals. The SOD content of various tissues is shown in table 5.1.

The inducibility of SOD by oxygen has been demonstrated in anaerobic and microaerobic bacteria. The ability of *Micrococcus radiodurans* to resist the effects of ionizing radiation has been attributed to its great content of SOD. Treatment of solid tumors with anthracyclin antibiotics (Adryamycin, Daunomycin) results in an increase in the content of SOD. These drugs act by producing $O_2^\bullet$ and other reactive oxygen species during their metabolism. Increased MnSOD has been observed in rats irradiated with sublethal doses of X-rays [213].

Chronic exposure of volunteers to a sublethal dose of ozone (0.7 ppm) induced an increase in SOD and glutathione peroxidase in the blood as an adaptive response. The same response was also observed in rats exposed to hyperbaric conditions and in humans exposed to tobacco smoke and paraquat

Table 5.1
SOD content of tissues in humans. After Marklund [177]

Tissue	Extracellular	CuZn	Mn
	expressed in units / gram net weight		
Adrenal	180	22,000	1,200
Brain, gray matter	80	23,000	900
Brain, white matter	120	11,000	280
Duodenum	350	8,000	310
Heart	210	12,000	1,300
Kidney, cortex	232	30,000	1,800
Liver	60	96,000	2,300
Lungs	650	7,600	380
Pancreas	1,650	10,000	800
Uterus	1,750	8,100	100
Spleen	100	13,000	320

intoxication.

These experimental observations suggest there may be a therapeutic opportunity to treat some conditions with purified SOD. Such treatment, administered in liposomes or albumin coated beads, has produced some improvement in cases of chemical intoxication, cataract, and rheumatoid arthritis. The efficiency of such treatment depends on the ability of the enzyme to penetrate to the affected site.

Some lipophilic, synthetic SODs have been proposed. These are mainly organic complexes of copper or zinc, and are of more scientific than practical interest.

5.3 GLUTATHIONE

This tripeptide (L-glutamyl-L-cysteinyl-glycine), with its dependent enzymes, is a central component of the nonspecific antioxidant defense system. In its reduced form (GSH), it is the main carrier for reactive sulfhydryl (SH) groups. Due to its strong nucleophilic character, GSH will react with many substances possessing electophilic centers, such as the important chemical pollutant, benzylchloride (Ph-CH$_2$Cl):

$$Ph\text{-}CH_2Cl \rightarrow Ph\text{-}CH_2^{\bullet} \xrightarrow{\ GSH\ } Ph\text{-}CH_2\text{-}SG + H^+ + Cl^- \qquad (3)$$

In many cases, reaction (3) proceeds nonenzymatically, but often it is catalyzed by GSH transferase in order to form mercapturic acids, which can be excreted in the urine.

Mixed disulfides between GSH and other thiol containing proteins have also been observed. GSH functions to form or maintain protein thiol groups, which may be required for catalytic activity or be involved in protein assembly or degradation.

An overall scheme of GSH metabolism is presented in figure 5.2. In addition, GSH is required for maintaining the integrity of the membrane structure, protein conformation, and biosynthesis of DNA and RNA.

The great concentrations on GSH in liver and other organs (1 - 15 mM) are sufficient for these requirements. However, when an acute intoxication occurs, there is a general drop in GSH concentration as GSH is used in the synthesis of mercapturic acids. A critical threshold in the intracellular concentration of GSH has been observed. Below this threshold, lipid peroxide formation increases [53, 182, 183].

The decrease in GSH in various affected organs is a dynamic process that is reversible, provided the chemical pollutant does not inhibit the enzyme, glutathione reductase, necessary for the regeneration of GSH. If this enzyme is inhibited, oxidized glutathione (GSSG) accumulates and is removed from the cells by an ATP-dependent translocase found in the cell membrane. Because of this, the measurement of GSSG in the blood gives information concerning the degree of exposure to a toxic agent [177, 181].

Due to its multiple activities, GSH has a variable turnover rate, depending on the specific organ and the intensity of metabolic activity: 1 hour for kidney, but 5 - 7 days for erythrocytes, brain, spleen, and lungs. As shown in figure 5.2, the biosynthesis of GSH is a dynamic process in which the conditions of demand (oxidative stress) and the availability of cysteine and methionine are limiting factors [40, 182].

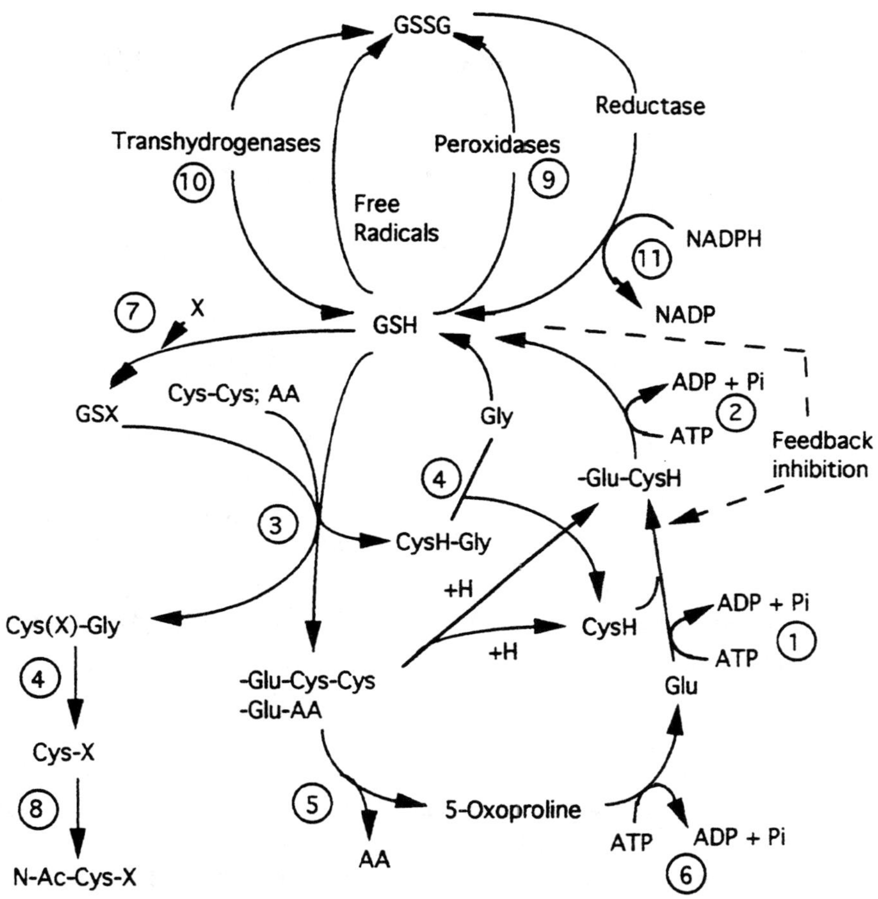

Figure 5.2. The biochemistry of glutathione (GSH). AA; amino acids. X;
compounds that form conjugates with GSH. Numbers indicate enzymes:
1. γ-glutamylcysteine synthetase, 2. GSH synthetase, 3. γ-glutamyltraspeptidase,
4. dipeptidases, 5. γ-glutamylcyclotransferase, 6. 5-oxoprolinase, 7. GSH S-
transferase, 8. N-acetyltransferase, 9. GSH peroxidase, 10. GSH thioltransferase,
11. glutathione disulfide reductase. After Meister [182].

As is the case for SOD, the therapeutic use of GSH has been studied for
the protection of organisms against ionizing radiation or chemical toxicity.
Molecules containing free SH groups are not stable due to their high
reactivity. So, for therapy, compounds that release free SH groups only *in
vivo* were tested. Only cystamine and N-acetylcysteine gave significant results
[23, 212, 266]. The experiments of Wright with GSH trapped in liposomes
showed some protective effect against the hepatotoxicity of paracetamol.

The high toxicity of methanol (blindness occurs with administration of only 7 ml) was attributed to its metabolic conversion to formaldehyde, which oxidizes all available SH groups, including GSH. The effects produced by the injection of a normally lethal dose of methanol (9 g/kg) or formaldehyde (0.14 g/kg) were reversed by the protective action of cysteine (38 mg), 2-mercaptoethanol (27 mg), or 2,3-dimercaptopropanol (known as BAL; British Anti-Lewisite) (4 mg). BAL is used successfully in the treatment of intoxication by As, Hg, and Au [40, 86, 287].

5.4 SELENIUM AND GLUTATHIONE PEROXIDASE

Selenium is a micronutrient, required for health when supplied in the range of 0.04 - 0.10 ppm. Above 4 ppm selenium is toxic. In man, selenium toxicity is rare. However, in some geographic areas, selenium toxicity in grazing animals is a serious problem, resulting in loss of appetite and weight. Recent data published by J. Neve, University of Brussels [204], has demonstrated selenium deficiency in man, especially in regions of China, New Zealand, and some African countries. The conditions in which selenium deficiency may play a role are listed in table 5.2. Some of these conditions may be improved by administration of selenium.

The protective effect of selenium in Cd, Hg, Pb, and Te intoxication is not clear since there seems to be a protective effect, even though the level of these elements in the tissue is not reduced. Selenium protection is noted in intoxication with paraquat and benzene. Several studies [86, 204] have demonstrated that a diet rich in selenium increases plasma HDL and erythrocyte GSH peroxidase levels.

During intoxication with metals such as Pb, or Cu, selenium seems to modify the binding of these metals to tissue proteins, keeping the free metal concentration in the cells high. With the free concentration in the cells high, the equilibrium between the cells and blood maintains a high level in the blood, also. By keeping the metals in the blood, they can be removed by the kidneys.

Table 5.2
Diseases and conditions in which selenium deficiency is a factor

Dietary deficiency
        Geography: China, Finland, New Zealand, etc.
        Premature new borns, elderly
        Pregnancy
        Vegetarianism
        Malnutrition
Gastrointestinal and kidney diseases
        Crohn's disease
        Mucoviscidosis
        Primitive biliary cirrhosis
        Alcoholism
        Kidney failure
        Hemodialysis
Muscular dystrophy, congestive cardiomyopathy, Keshan's disease, cetoid
lipofuscinosis, trysomy 21
Miscellaneous
        Arthritis
        Kaschin-Beck's disease
        Psoriasis
        Cancer
        Cardiovascular disease

Selenium is metabolized as dimethyl and trimethyl selenides and selenomethionine. Selenium also binds easily to free SH groups of GSH or proteins. The primary active form of selenium is glutathione peroxidase (EC 1.11.1.9), which is active in the decomposition of H_2O_2 and organic peroxides (fig. 5.1B). The enzyme weighs 90,000 and has 4 subunits, each containing 1 atom of selenium. There is also a form of GSH peroxidase that does not contain selenium. This enzyme has both peroxidase and transferase activity. Glutathione peroxidases catalyze the decomposition of a large range of peroxides derived from lipids (PUFAs), steroids (cholesterol, progesterone), nucleic acids (thymine), prostaglandins (PGG_2), and H_2O_2 [12, 75].

Glutathione peroxidase is very involved with the body's defense by catalyzing the decomposition of peroxides, having the overall reaction of:

$$2GSH + RH\text{-}OOH \xrightarrow{\textit{GSH peroxidase}} GSSG + H_2O + ROH \qquad (4)$$

GSH peroxidase competes with catalase for the decomposition of H_2O_2. In erythrocytes, both enzymes are found in the cytosol where they neutralize the reactive oxygen species and H_2O_2 that are readily formed there. In other cells, only one of the two enzymes is normally found to be present at high concentration. The possible role of selenium in protecting the body from chemical carcinogens is due to the presence of this atom in GSH peroxidase. The lungs of smokers contain about 40% more GSH peroxidase than nonsmokers, which is believed to be an adaptive effect.

The most readily assimilated form of selenium appears to be Se-yeast, which is available commercially from several companies (75 - 125 µg/day). In Germany, a GSH-peroxidase like drug called Ebselen is sold as an effective antioxidant.

5.5 ASCORBIC ACID

Ascorbic acid (vitamin C) is a widely discussed antioxidant that some enthusiasts have recommended should be taken in multiple gram amounts (RDA is 60 mg per day in adults). The value of greatly increasing the intake of ascorbic acid because of its importance in detoxification, antioxidation, and immunologic roles is controversial, but has not been disproven. It seems that any excess in intake of ascorbate is simply eliminated.

In the body, ascorbic acid is found equilibrated between its oxidized form (dehydroascorbic acid, DHA) and its reduced form (AH^-) with the reduced form predominate. Ascorbate readily reacts with superoxide ($O_2^\bullet$), having a rate constant of 2×10^5 mols/s, forming the ascorbate radical ($A^\bullet$):

$$AH^- + O_2^\bullet \rightarrow A^\bullet + H_2O_2 \tag{5}$$

This reaction is inhibited by SOD. There is abundant evidence that vitamin C is involved with the biotransformation of xenobiotics. With exposure to several chemical pollutants, such as paraquat, ascorbate is protective, but for electrophilic compounds, such as nitro derivatives (some of which are potential carcinogens), a new free radical semiquinone is formed:

$$AH^- + R\text{-}NO_2 \rightarrow A^\bullet + R\text{-}NO_2^\bullet + H^+ \tag{6}$$

This radical, $R\text{-}NO_2^\bullet$, rapidly reacts ($K = 108$ mols/s) with other compounds:

$$2R\text{-}NO_2^\bullet + 2H^+ \rightarrow R\text{-}NO_2 + H_2O + R\text{-}N{=}O \tag{7}$$

$$R\text{-}N{=}O + AH^- \rightarrow R\text{-}N^\bullet OH + A^\bullet + H^+ \text{ (nitroso)} \tag{8}$$

$$R\text{-}N^\bullet OH + O_2 \rightarrow R\text{-}N{=}O + O_2^\bullet \tag{9}$$

Consequently, ascorbic acid may be coupled with reactions that favor the formation of free radicals. This may provide an explanation of why ascorbic acid in low concentrations can act as a pro-oxidant, favoring peroxidation in tissues, especially when Fe^{2+} is also present. Ascorbic acid is most effective as an antioxidant at high concentrations.

In addition to its ability to react with oxygen radicals, ascorbate can react with $OH^\bullet$:

$$OH^\bullet + AH^- \rightarrow A^\bullet + H_2O \tag{10}$$

$$A^\bullet + A^\bullet \rightarrow AH^- + DHA^- \tag{11}$$

Consequently, ascorbic acid can neutralize all reactive oxygen species, including the hydroxyl radical, if conditions are right. The regeneration of ascorbate occurs through reaction (11) or through the reaction with GSH (figure 5.1).

Ascorbate can be detected in plasma by chemical determinations as well as through electron spin resonance (ESR), but the utility of the later technique is still under debate [21, 125, 161].

Ascorbic acid functions as a physiologic antioxidant because the semihydroascorbate radical ($A^\bullet$) has a low reactivity. Consequently, the radical is trapped and cannot be passed along. Enzyme systems exist to reduce this radical back to ascorbate at the expense of NADH and glutathione reductases. Since these enzymes are largely intracellular in location, vitamin C is rapidly depleted in extracellular spaces during oxidant stress. Ascorbate is normally present at a concentration higher than other antioxidants. In liver the level of vitamin C is about 2×10^{-3} M, while vitamin E is only present at about 2×10^{-5} M. Under *in vitro* conditions of oxidant stress, ascorbate is the first consumed of all the water soluble antioxidants. Therefore, a higher intake of vitamin C has been recommended under stressful conditions or exposure to chemical pollutants, provided it is adequately buffered.

5.6 OTHER NATURAL ANTIOXIDANTS

Vitamin E, or α-tocopherol, is a known antioxidant that has been of increased clinical interest in recent years. Twenty years ago clinicians referred to vitamin E as "a vitamin in search of a disease" because deficiencies in man could not be found. The great power of vitamin E as an antioxidant is its lipid solubility and capacity of inhibiting all the steps of peroxidation. It is carried in low density lipoprotein, which assures the vitamin entry into cells because of the cell LDL receptor protein.

Vitamin E, because it is lipid soluble, is in close contact with the lipid membrane and lipid metabolism [211, 244, 278]. This relationship places the antioxidant vitamin close to the potential activity of xenobiotics, many of which react with lipids (insecticides, PAH, PCB, etc.). Its lipid solubility allows vitamin E to readily cross the blood-brain barrier, increasing the otherwise low level of antioxidants in the brain [56, 83, 108]. Vitamin E also increases biliary flow, decreases plasma cholesterol and reduces the formation of biliary stones. There is evidence the antioxidative effects of vitamin E provide some protection from athersclerosis [53, 181]. Due to its lipid solubility, vitamin E acts mainly at the level of cell membranes by scavenging peroxyl radicals, preventing them from participating in chain-propagating steps. The normal intake of vitamin E ranges between 4 and 22 IU/dl (international units per deciliter) in adults, maintaining an average blood level of about 23 μmol/L.

Carotenoids are widely distributed in nature where they play an important role in protecting cells and organisms. Carotenoids, such as lycopene and β-carotene, can inactivate electronically excited molecules like singlet oxygen ($^1O_2$) by a process called quenching. Singlet oxygen is generated by photochemical reactions and in some enzymatic peroxidation reactions of biomembranes. The antioxidant property of carotenoids may be a possible explanation of the role of this molecule in the prevention of cancer, since β-carotene may have anticarcinogenic properties that are independent of its provitamin A activity. An inverse relationship between β-carotene intake and the incidence of certain types of cancer, including lung and gastrointestinal tract cancers, has been reported [68, 83]. The role of carotenoids in protecting plants against photo-effects induced by their own chlorophyll, as well as in the treatment of patients with photosensitivity diseases, sun burn is clearly established, but the absolute mechanism by which β-carotene may be protective against cancer is still unknown.

Ceruloplasmin, or ferroxidase (EC 1.16.3.1), is a protein containing 8 atoms of copper per molecule and is the main plasma oxidase. Its substrate is supposed to be Fe^{2+}, but it can oxidize a range of compounds, including catecholamines, tyrosine, diamines, aminophenols, phenothiasines, etc. In addition to its detoxification ability, ceruloplasmin possesses dismutase and antioxidative properties. Ceruloplasmin also acts on the immune system by participating in the process of opsonization of pathogens. Thus, ceruloplasmin belongs to the acute phase proteins, a family of proteins whose levels are greatly increased during infections and inflammatory conditions as well as in chemical intoxication. Hepatotoxic compounds, such as carbon tetrachloride, cause a decrease in plasma ceruloplasmin by damaging the parenchymal tissue [107]. In normal plasma, the ratio of ascorbate to ceruloplasmin is approximately 1:13 to 1:65, ensuring a high level of ferroxidase activity. When the ratio of ascorbate to ceruloplasmin is greater than 100:1, the feroxidase activity of ceruloplasmin becomes inhibited. This effect may be a mechanism to assure the sequestration of reactive ferrous ions.

Zinc is a cofactor for nearly 80 enzymes, including carbonic anhydrase. This element activates the function of insulin, vitamins B_1 and A, thyroid stimulating and luteotropic hormones, and the biosynthesis of DNA. Zinc is mainly found in the erythrocyte (85%) and plasma where it is reversibly bound to albumin, a-macroglobulin. Its exact level varies with age and sex.

The antioxidative activity of zinc seems to be mainly due to its role in stabilizing membrane structures. It has been used as an antidote in heavy metal intoxication and in some drug exposures (etanbutol, disulfiram, isoniazide, and penicillamin). The blood level of zinc decreases in many pathologic conditions such as infections, surgical trauma, cancer, liver or pancreatic failure, renal dialysis, etc. Its blood level also decreases after a diet rich in cereals (high in phytic acid), sunflower seeds, and red wine. Zinc favors wound healing and possesses a high, unexplained, immunologic activity. Its presence at a suitable level in the diet is protective against chemical intoxications.

Uric acid is the most recently discovered antioxidant in the body. This activity of uric acid discovered by Ames [6, 7]. Uric acid is the final product of purine catabolism. *In vitro*, it inhibits the peroxidation of PUFAs, quenches oxygen, and protects nucleic acids, erythrocytes, and hyaluronic acid from reactive oxygen species. In addition, uric acid activates the synthesis of prostaglandins.

Other molecules with antioxidant or suspected antioxidant properties include:

- Albumin is able to bind a range of metallic ions, bilirubin, aromatic and compounds making them unavailable for peroxidative reactions. It is present in large amounts in the plasma, giving it a large antioxidative capacity.
- Metallothioneins are a specific class of proteins that are able to bind many metal ions (especially Cd, Cu, and Hg).
- Some amino acids, such as histidine and taurine, are able to scavenge oxygen and covalently bind ions of heavy metals (Cd, Hg).
- Flavonoids, polyphenols, and tannins from the diet are also potent antioxidants, acting directly against oxygen activation or by binding heavy metal ions. The so-called "French paradox" involves the beneficial effect of a daily glass of red wine against cardiovascular disease. This protective effect has been demonstrated epidemiologically. The French habit of a glass of red wine a day correlated with a low incidence of cardiovascular disease. This action is due to the high content of flavonoids and polyphenols in the wine.

PART III: WHEN THE DEFENSES FAIL

CHAPTER 6

ENVIRONMENTALLY CAUSED DISEASES

U p until some 20 years ago, the suggestion that environmental factors, other than occupational exposures, could be the cause of serious disease would have been met with skepticism. Now there is considerable evidence that all people are susceptible to varying extents to diseases caused by pollution. Given the huge amount of waste released to the environment it would be surprising if there were no effect on people's health. For example, in the United States 145 million tons of smoke and dust, 400 million tons of solid wastes, 600 tons of hot waste water, 50 million tons of used bottles and tin cans, and 7 million used cars [45, 183]. Chemical pollutants are not broken out from this list, but constitute a significant portion.

Part of the reason exposure to chemical pollutants can cause a problem is the ability of organisms to accumulate pollutants. The accumulation of pollutants leads to a situation in which the local environmental level may not be toxic, but the level present in the organisms living there builds up to the point it is toxic. Thus, the overall effects of pollution accumulating in an organism may be like the iceberg, only the tip can be seen (Fig. 6.1). At point 2 on figure 6.1, some clinical signs of toxicity may appear, but frequently go unobserved. This is particularly unfortunate since the effects at this point are mostly reversible. The appearance of obvious clinical signs of illness does not reliably occur until the toxicity threshold is exceeded (points 3 and 4) [92, 121].

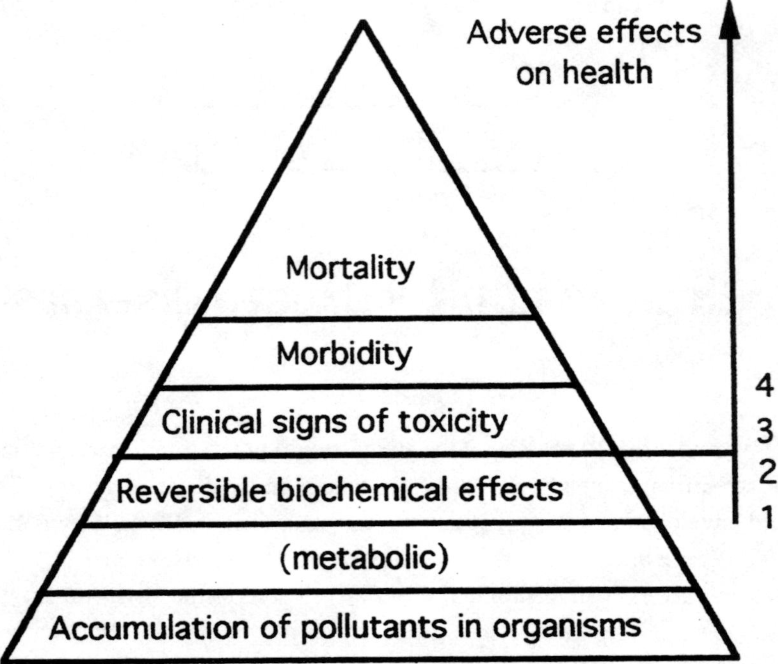

Figure 6.1. The ralation between the accumulation of chemical pollutants and the appearence of clinical symptoms of toxicity: 1 - modification at the molecular level, 2 - metabolic alteration, 3 - transient clinical symptoms, 4 - overt clinical symptoms.

We now know that even a single exposure to some pollutants can eventually lead to adverse effects, even though there initially seems to be no effect. Thus, a month may pass between a single exposure to the herbicide Paraquat and the development of gastrointestinal and lung effects. The lung effect is caused by the proliferation of fibroblast cells at the inner surface of the lung (pulmonary epithelium) reducing the ability of the lung to oxygenate the blood. This effect may be fatal. Some organophosphate insecticides are neurotoxins that can act several weeks after exposure. The resulting paralysis is the consequence of neurological damage in the brain. Such pollutants, called "hit and run poisons," often act without leaving biochemical traces [63].

The biological consequences of exposure to chemical pollutants can be acute or chronic, depending on the mode of exposure, pollutant concentration, general health and genetic factors of the organism, as well as dietary and social factors. All of these factors are in force at once, and may result in effects that are insignificant, temporary, or life threatening. The way by which a pollutant enters the body can cause the major effect to be localized in the lungs, the liver, or some other part of the body. The biological effects of other compounds, such as carcinogens or heavy metals are independent of the means of entry, but may be affected by the other factors mentioned above [11, 233].

The central nervous system plays a role in the adaptation of the body to environmental factors (fig. 6.2). Consequently, many of the symptoms reported in people who claim to be suffering from an ecological disease are psychological or neurologic in origin. Many of these illnesses involve one or more organs that exhibit sensitivity to the environmental agent. Common sites are the nose, upper respiratory tract, eyes, and gall bladder. In some cases there may be a poorly defined allergic response. The hypersensitivity involved in these reactions has wide variations [120, 153].

Commonly, adverse reactions to chemical compounds have been divided into two categories: toxic reactions, where the agent presumably interacts with organs and tissues, and sensitivity or hypersensitivity states, where an immune reaction may be involved (allergies).

Toxic agents produce diseases that:

- Exhibit distinct histological lesions in one or more organ systems
- Involve dose dependent tissue modifications
- Affect all individuals in a given population
- Are reproducible in experimental animals
- Appear predictable and with a brief lag period

Sensitivity and hypersensitivity states to certain chemical usually:

- Produce variable pathological changes, both in the extent of lesions and in the time of appearance
- Exhibit no definite dose-dependent relationship, either in the occurrence or severity of lesions
- Affect a small number of individuals in a given population
- Are not readily reproducible in experimental animals

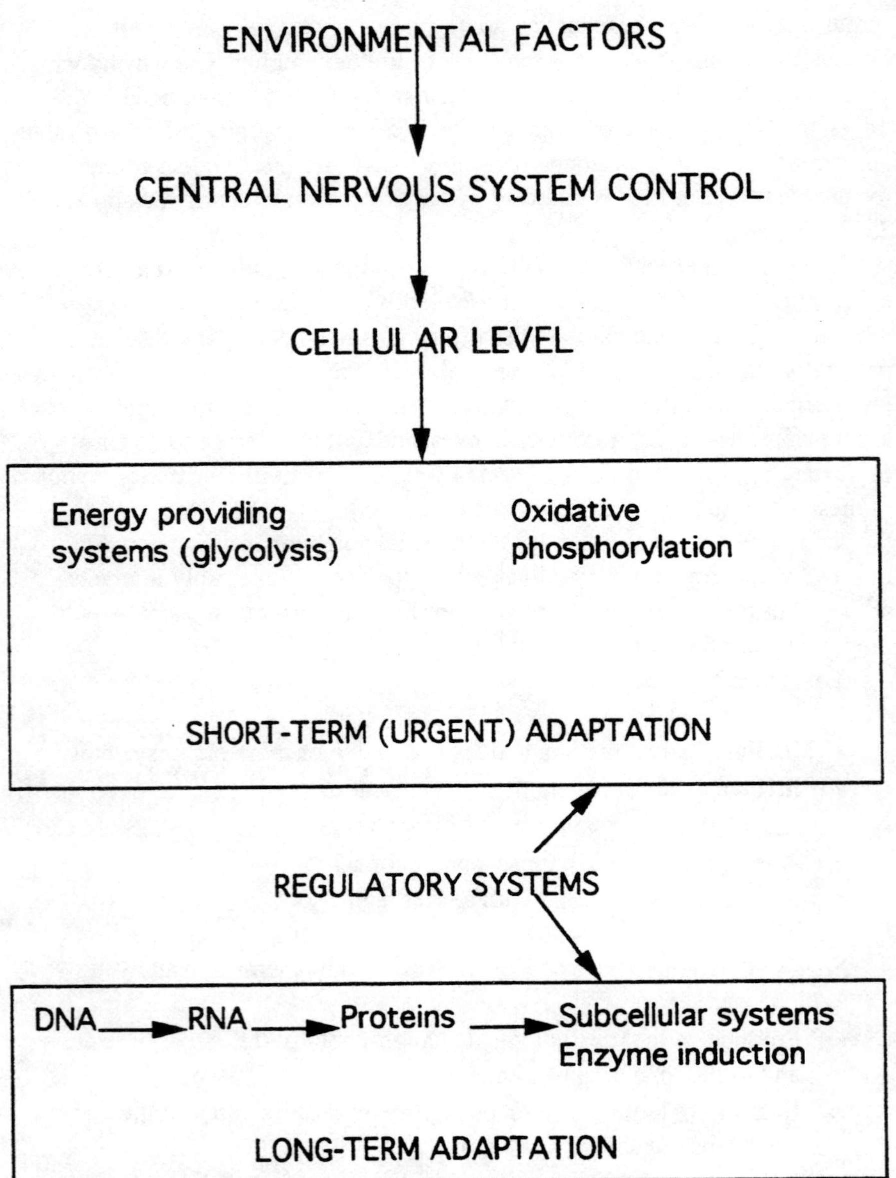

Figure 6.2. Short - and long-term adaptations. After Meerson [183].

This later group of reactions reflects the individual susceptibility of a person, relating to an immune reaction. Many of these reactions are related to the bonding of the agent to cellular macromolecules, forming haptenic antigens and the classic involvement of IgE (immunoglobulin E).

In addition to these two groups of reactions to chemical compounds, there now appears to exist a third, controversial, reaction. A significant number of people claim to suffer clinical reactions to low levels of chemical compounds commonly found in the environment, including the home and place of employment. This "chemical sensitivity" disease remains controversial because biochemical or biological modifications sufficient to explain the disease have not been found. However, as many scientists argue, millions of people are affected and exposure to a variety of chemicals is widespread and will continue.

In theory, four groups of people could be expected to show chemical sensitivity:

- People chronically exposed to chemicals, such as certain industrial workers and miners
- Occupants of tight, energy efficient buildings, including office workers and schoolchildren
- Residents of communities whose air is contaminated by chemicals
- Individuals with direct exposure to various chemicals that can be found in domestic indoor air: pesticides, plastics, etc.

In spite of great individual variation (age, sex, education, nature of chemical, level of exposure) the reported symptoms frequently involve multiple organs with variation in the nature and severity of the symptoms. Most of these symptoms seem to be very subjective, involving the central nervous system. These symptoms include chronic fatigue, irritability, and difficulty in concentrating. Generally, hematological and biochemical tests show no abnormal results.

People suffering from chemical sensitivity often exhibit non-classical or non-allergic symptoms. These symptoms seem to originate from some acute exposure to the chemical, after which the triggering of a response occurs at very low levels of chemical exposure. The initial chemical need not be the same as the chemical that later causes a reaction. Sometimes the initial agent is described as "sensitizing" the individual who is then referred to a being sensitized.

Unlike classical toxicity, the effect of low level exposure to chemicals is not the same as that observed in normal populations at higher doses. The lack of symptoms in most people, even at higher doses, partly explains the reluctance to accept chemical sensitivity as a genuine illness. However, in 1987, the National Research Council in the United States estimated that approximately 15% of the population may experience "increased allergic sensitivity" to chemicals. This allergic sensitivity is not mediated by IgE mechanisms. The proportion of the population affected by chemical sensitivity seems to be increasing and could become a major problem for some geographical areas.

The problem of chemical sensitivity is linked to the complex concept of adaptation. Mankind has been found to be able to tolerate or adapt to an enormous range of substances, including: 1) outdoor and indoor pollutants, both at home and in the workplace, 2) food and water additives and contaminants, 3) drugs, cosmetics and other consumer products. History indicates that as a country has become more industrialized (or civilized), its citizens have been exposed to a greater variety and greater quantities of chemicals. Adaptation allows the population to accept exposure to these chemicals without suffering ill effects. However, the reverse can also be true. In some instances, chemical exposure can have an adverse effect, leading to diseases or symptoms of chemical sensitivity.

Clinical ecologists and allergists point out that for each chemical compound there is a threshold or a total exposure level, which must be passed before symptoms appear. There is variation in individual's ability to respond to chemical stress through the production of antioxidant enzymes, etc. In some people these pathways may be defective, so that the body fails to properly metabolize and dispose of some chemicals. Consequently, some people are more susceptible to chemicals in food and the environment that, because they are not removed by the protective systems, cause chemical sensitivity. These people may show signs of chemical sensitivity when exposed to very low levels that are tolerable to the rest of the population.

Table 6.1
Relative severity of symptoms of illnesses
with possible environmental causes

Symptom	Chemical sensitivity	Chronic fatigue syndrome	Sick building syndrome
Chronic fatigue	++	+++	+
Psychological disorders	++	++	++
Headache	++	++	++
Allergic rhinitis	++	+	+++
Dermatitis	++	+	+
Arthritis	++	+	+
Infections	Candida albicans	flu	pneumonia

Psychiatrists have expressed the opinion that people who claim to be suffering from chemical sensitivity are prone to atypical depression, hypochondriac behavior, post-traumatic stress syndrome, hysteria, panic disorder, or a combination of these problems. If chemical exposure can directly effect the central nervous system, then it may fairly be asked if these people experience chemical sensitivity because of a psychological disorder, or if the psychological disorder is a result of chemical sensitivity.

Since there is reason to believe the central nervous system is involved in chemical sensitivity, it is interesting to compare the symptoms associated with chemical sensitivity and some controversial illnesses. The symptoms in the first column of table 6.1 were reported by W. I. Rea, et al. [233] and involved evidence of immune dysfunction (depressed T lymphocytes). In chronic fatigue syndrome, a reduced CD8 suppresser cell population and increased activation of CD38 markers were observed. While in this specific study, a viral-like agent was involved, in the other references the symptoms followed chemical exposure.

Dr. Black [27] suggests there are two subsets of people who suffer from environmentally related diseases: those with true physical disorders and those with true psychological disorders. Dr. Miller [11] indicates sensitivity reactions to low levels of chemical exposure may involve the limbic system (a part of the brain involved in autonomic functions and certain aspects of emotional behavior), which secondarily can affect the immune system and then different organs in a variety of ways. Chemical sensitivity has also been suggested to result from an altered helper to suppressor T-cell ratio and an altered production of B-cells (lymphocytes).

Children seem to be more sensitive to environmental chemicals than adults. Atopic dermatitis is a form of eczema, that is most prevalent in children under five years of age, and is characterized by dry, easily irritated skin with intermittent flare-ups of eczema. A variety of triggers are known to initiate flare-ups of eczema, including irritants (soaps and other cleaning chemicals), heat, and humidity. The prevalence of atopic dermatitis has increased over the past several decades and now affects 10 - 15% of the population at some time during their lifetime [233]. Like asthma, atopic dermatitis can be divided into extrinsic and intrinsic forms of the disorder. Patients with extrinsic atopic dermatitis are generally younger and have significant allergen-specific IgE antibodies. The are sensitive to food and/or airborne allergens that trigger flares of eczema. Patients with intrinsic atopic dermatitis are generally over 20 years of age and have no evidence of allergen-induced flare-ups of eczema. Acute lesions are characterized by edema and intracellular infiltration of leukocytes (mostly lymphocytes).

These $CD4^+$ lymphocytes are antigen-specific TH_2 cells, which, when activated, secrete cytokines (IL-4, IL-10, and GM-CSF). These cytokines promote B lymphocyte production of IgE and IL-1 receptors on monocytes. A consistent increase in cyclic-AMP phosphodiesterase activity is also observed in atopic dermatitis patients.

Atopic dermatitis patients experience greater susceptibility to skin infections, commonly *Staphylococcus aureus* and Herpes simplex virus. Patients with atopic dermatitis and food allergies have an elevated spontaneous basophile histamine release (SBHR) from peripheral blood basophiles. The ingestion of food allergens appears to provoke the histamine release, which could lower the threshold of activation for mast cells. Similarly, an association between bronchial hyperactivity in patients with asthma and SBHR by peripheral blood monocytes is reported. Inhaled allergens (pollen, mold, dust mites, or mineral dust) absorbed by the respiratory mucosa exhibit effects at distal skin sites.

Patients often comment that anxiety, anger, and frustration trigger the itching and flare-ups common in atopic dermatitis. Stressful conditions and negative emotions may trigger latent hypersensitivity of atopic dermatitis.

Chemical pollutants can also be involved in this hypersensitization as illustrated by the reported symptoms of chronic fatigue syndrome (Table 6.1).

Rhinitis is the most common symptom for environmental illnesses reported in table 6.1. Rhinitis is not actually a disease, but is nasal irritation or inflammation. Symptoms of rhinitis include: runny nose (rhinorrhea), itching (puritus), sneezing, and nasal congestion. These symptoms are the nose's natural response to irritation caused by allergens, chemical exposure (including cigarette smoke and irritant gases), temperature changes, infections, etc. [202, 233].

The nose normally produces mucus, which traps invading substances like dust, pollen, chemical irritants, and bacteria. The mucus containing these substances is shed periodically. Nasal congestion is a natural response to irritation and inflammation. However, severe congestion can result in facial pressure and pain, as well as dark circles under the eyes.

Sinusitis is not the same as rhinitis, but is the inflammation or infection of the sinus cavities, which are found in the bone of the skull and open onto the nasal passage. The symptoms of allergic rhinitis (hay fever) are caused by exposure to different substances (allergens) to which the patient may become sensitive (develop an allergy). Common allergens are tree, grass, and weed pollen; mold; animal hair, and house dust mites. In addition to allergic reactions, congestion is caused by the common cold, which can result from exposure to one of over 200 different viruses. Another interesting, but poorly understood type of rhinitis is the vasomotor rhinitis [193, 233].

Many people who have never been diagnosed with an allergy complain of recurrent nasal congestion, runny nose, itching, and other symptoms suggestive of allergy. The triggers of this vasomotor rhinitis include cigarette smoke, strong odors and fumes, laundry detergents, cleaning solvents, chlorine in swimming pool water, car exhaust, and other air pollutants. Vasomotor rhinitis is aggravated by spices used in cooking, alcoholic beverages (particularly beer and wine), aspirin, and certain blood pressure lowering drugs (such as Reserpine and Propanolol). Some people are unusually sensitive to abrupt changes in the weather or temperature. As with chemical pollutants, these agents do not usually induce the formation of allergic antibodies and do not produce positive skin test reactions. Some people are unusually sensitive to irritants, developing significant nasal symptoms when exposed to even low chemical concentrations. Thus, the vasomotor response seems to be an exaggeration of the normal nasal response to irritation, occurring at levels of exposure that don't bother most people. It

Table 6.2
Common uses of "Allergen"

Use of "Allergen"	Example	Comments
An organism that produces allergic molecules.	Ragweed	Several species producing common and unique allergenic molecules.
A particle bearing allergenic molecules.	Ragweed pollen	Morphologicaly similar forms with common allergenic molecules.
An aqueous extract of allergen bearing material.	Short ragweed extract	Mixed macromolecules, some of which are allergenic.
A molecule that stimulates the production of and that binds to IgE antibody.	Ambal	A specific molecule with one or more epitopes recognized by human IgE.

also occurs more often in smokers and older individuals [191, 202].

As is the case with allergic rhinitis, vasomotor response often cannot be cured, but the symptoms can be kept under control by avoiding exposure to the triggering agent and by taking medications (antihistamines, cromolyn, etc.).

The diagnosis and treatment of allergic diseases depends on an accurate understanding of the relationship between allergen exposure and the resulting illness. As shown in table 6.2, the term "allergen" has been used in a variety of connotations.

According to the dose/response relationship, the molecule that actually induces the allergic response is the true exposure unit. But, clinical evidence shows that if a person becomes sensitized to allergens from one source, the patient may react to the same or similar allergens from different sources. This is a consequence of cross-reactivity at the molecular level, where groups of antibodies are able to recognize more than one binding site (epitopes) on a molecule. Dual sensitization occurs from exposure to two or more allergens that share an environment (e.g., dust and mites). The results and interpretation of allergic skin tests are often very subjective, depending on the way the tests are done and the standards used for interpretation [191, 267].

A correct diagnosis of chemical sensitivity is made much more difficult by the extreme individual variation that occurs in people. In order to make a diagnosis, specialists have patients avoid exposure to the suspected triggers, including fasting if necessary, for 4 - 7 days. Then individual foods are reintroduced, one per meal, followed by use of local drinking water, and finally the patient is challenged with very low levels of common chemicals.

The problem of chemical sensitivity in all its varied forms suggests some people possess a certain hypersensitivity to some chemicals, commonly chemical pollutants. The majority of chemical pollutants are metabolized in the endoplasmic reticulum of the liver, involving cytochrome P_{450}. As discussed elsewhere in this book, metabolism of chemical pollutants often leads to an increased production of reactive oxygen species as a side effect. We propose that reactive oxygen species are directly involved in chemical hypersensitivity. If reactive oxygen species are released during the metabolism of chemical pollutants, a mild oxidative stress should result because the amount of released reactive oxygen species would exceed the available antioxidant capacity of the cell. This could lead to some forms of biochemical hyperactivity. Some recent work supports this hypothesis. Bondy [30] presents evidence showing a direct relationship between excessive amounts of reactive oxygen species and neural hyperactivity, as well as between oxidative stress and various neurological disorders. Increased levels of glutamate are associated with allergic responses [289], and have recently been shown to also increase the formation of the hydroxy radical [301]. An antagonistic relationship is known to exist between glutamate, which favors oxidation, and the hormone melatonin, which has antioxidant properties. In spite of this favorable evidence, more studies are needed to firmly demonstrate this hypothesis.

CHAPTER 7

CARCINOGENESIS AND MUTATION

7.1 OVERVIEW

As previously mentioned, the most dangerous consequences of exposure to chemical pollutants are carcinogenesis, mutagenesis, and teratogenesis. These have been the subject of numerous scientific congresses [6, 110, 122, 123, 185, 281].

The first clinical report of an occupationally related cancer was reported 300 years ago in Jachymov, Czechoslovakia, where miners were observed to be prone to a pulmonary cancer, then called "mountain disease." As recently as 20 years ago, many scientists were unconvinced of the existence of a connection between environmental exposures and cancer [123]. However, as statistical and epidemiological data and the results of animal studies have accumulated, these doubts have ceased. In response to the better scientific understanding of cancer risks and changes in public opinion, protective measures have been taken in some countries. Today, even though a significant number of scientists believe 80 - 90% of the cases of cancer have a connection with chemical pollutants, it is difficult to support the claim that cancer is an "ecological disease." The evidence for and against this claim is:

a. We still have an imperfect understanding of the agents and mechanisms involved in the transformation of normal cells to a highly proliferative state;

b. Even in experimental studies, the incubation time for the appearance of clinical signs of cancer following chronic exposure to chemicals is long. In humans it is estimated to be 5 - 20 years. This long incubation period, along with individual variation, suggests the presence of defensive mechanisms that must be overcome before cancer can become clinically significant.

c. Today, an increased frequency of cancer is believed to parallel the development of industry. Many synthetic compounds are not recognized by the body's detoxification systems. The alleged involvement of chemical pollutants in carcinogenesis largely stems from the observation that certain cancers seem to be common in certain professions. In the nineteen-seventies the involvement of chemicals in specific cases of cancer was recognized:

- Metal dust and solvents in pulmonary cancer
- Vinylchloride in liver cancer
- Aromatic compounds (naphthylamine) in gall bladder cancer.

d. Animal tests on the carcinogenicity of chemicals can not always be extrapolated to humans. This applies both to the risk chemicals present and to the critical amount needed to cause cancer. There is no universally accepted animal test to estimate the carcinogenic threshold of chemicals.

Table 7.1 presents a list of recognized potentially carcinogenic compounds. These compounds promote the formation of cancers after entering the body and being metabolized. The mechanisms involved in this chemical-induced carcinogenesis are only partly known. Regardless of the biological, immunological, viral, or biochemical hypothesis of the cause of cell proliferation, the earliest steps involve the modification of DNA. The modification of DNA directly results in alterations in protein synthesis and cell division.

Several authors [7, 122, 166] define the DNA lesion as, "any modification that alters the code and normal function of transcription and replication." This definition can include a great variety of damaging agents, some of which are listed in table 7.2. Most of the mechanisms listed in this table are the result of the OH$^\bullet$ radical produced during oxygen activation. Note that some of the same agents are present in tables 7.1 and 7.2 (radiation, heavy metals, and aromatic compounds). Metabolic activation of some chemical pollutants may, directly or indirectly, alter other metabolic processes, favoring the production of endogenous mutagens. This effect has been observed to lead to increased S-adenosylmethionine (a methylating agent of DNA), N-nitrosoderivatives

Table 7.1
Potential carcinogens

Humans	Animals
Coal tar	4-aminobiophenyl
Mineral oils	Yperite
Cresol	Stilbesterol
Aromatic amines	Aflatoxin
Benzene	
Petroleum residues	
UV radiation	
Ionizing radiation	
Tobacco	
Heavy metals	
Asbestos	
Betel nut	

Table 7.2
Agents that may damage DNA

A. Chemical-induced transformation:	Examples
exogenous agents	- UV and ionizing radiation - mutagens from food - tobacco smoke - chemical pollutants
endogenous agents	- some drugs - alkylating agents (N-nitroso derivatives) - formaldehyde and other aldehydes
B. Mechanical-induced transformation:	
physicochemical scission of DNA strands	- enol or imine tautomerization - ionization - base substitution - hydrolysis and loss of bases - free radicals from purine bases - strand breaks (sugar moiety)

from gastrointestinal fermentation, and surprisingly high levels of formaldehyde in the blood (over 0.1 mM). Reactive oxygen substances are

Table 7.3

Magnitude of DNA damage following chronic and acute exposure to radiation. After Singh [262].

Event	Rate at background (150 mrad/year)	Rate at LD50 (4.5 Gy)
Number of spurs	$3X10^{-3}$ cell^{-1} hr^{-1}	$1.4X10^3$ cell^{-1} hr^{-1}
Number of spurs on DNA	$3X10^{-6}$ cell^{-1} hr^{-1}	11.4 cell^{-1} hr^{-1}
Number of expected DNA strand breaks	$3X10^{-6}$ cell^{-1} hr^{-1}	11 cell^{-1} hr^{-1}
Damage by organic compounds	1 per hour	1 per hour

among the most frequently formed compounds, acting either directly or indirectly.

Since ionizing radiation and chemical pollutants product nearly identical DNA damage, and OH$^\bullet$ is produced in both situations, it is worthwhile to look at the work of radiobiologists (Table 7.3). The assumptions made here are 1) an average human weight of 70 kg, 2) average cell volume of $4x10^{-12}$ cm^3, 3) average cell weight of $4x10^{-12}$ g, 4) average energy per spur of 60 eV, 5) one spur on DNA produces one double-strand break.

The magnitude of the effects on the DNA molecule depends on the amount of radiation produced free radicals. The number of double strand breaks is approximately proportional to the square of the free radical concentration. This high level of DNA damage following exposure to radiation or chemical pollutants will produce a large amount of altered proteins, unless there are sufficient antioxidants and repair enzymes to prevent it.

The uncontrolled cell division that is characteristic of cancer may be the result of an accumulation of DNA lesions over time. Generally, it is not possible to directly detect the altered DNA, so only the byproducts released in the blood and urine are easily available to provide information on the alterations taking place inside the cells. Studies on cells have found apparently normal cells containing a high number of chromosomal abnormalities and tumor cells with no detectable changes in chromosome number or structure [119, 123, 166]. Chemical pollutants with carcinogenic properties may also act at the level of transfer RNA. Hybridization studies on nucleic acids showed that, in tumors, messenger RNA is little altered (some 35 nucleotide sequences out of at total of 10,000). It is possible that a single sequence is responsible for the alteration of the entire cell division process.

7.2 EXPOSURE TO POTENTIAL CARCINOGENS

A long period of exposure to chemical pollutants is required to overcome the natural protective systems of the body. Consequently, chronic long-term exposure to a carcinogenic chemical pollutant is likely to lead to cancer. Just how long and how extreme an exposure is required to lead to cancer is one of the most important activities of the International Agency for Cancer Research (IACR), Lyon, France. The data we will describe here was collected by the IACR in Finland and examined the effect of a large number of potentially carcinogenic chemical pollutants. Finland was selected for the study because of its small population (5 million inhabitants) and high level of industrialization. Only chemical pollutants with potential carcinogenic properties that had been demonstrated in animals were considered. The conclusions obtained from such a study can be extrapolated to other countries of similar demography.

The interesting conclusions from the Finland study are presented in table 7.4. First, a great variation in the exposure to chemical pollutants is seen, ranging from 0.1 to 65,000 μg/day. Second, heavy metals, notably cadmium, are found in marked amounts in food, air, and water. Third, polyaromatic hydrocarbons (PAH) are present in all tested samples: 2 μg in food, 0.2 μg in water, 0.4 μg air, and 1 μg in tobacco smoke. The carcinogenic properties of PAH have been well documented [110, 119, 272].

As mentioned earlier, a significant number of chemical pollutants are potential carcinogens or mutagens for animals. Included are compounds with a great variation in structure: nitrates, flavonoids, organophosphorylated pesticides, acetaldehyde, etc. Except for pesticides, these compounds are present in the Finland samples in amounts as high as tens of milligrams.

Table 7.4
Exposure to potential carcinogens in the environment [122].

Source	Compound	Exposure (μg/day)
FOOD:		
Plant components	Hydrozines	1
	Flavonoids	50,000
Metals	Arsenic	60
	Cadmium	13
	Chrome	30
	Lead	70
	Nickel	130
Organic compounds	Chlorinated phenols	5
	DDT derivatives	2
	Hexachlorobenzene	0.5
	Polychlor-biphenyls	7
	Polycyclic hydrocarbons	2
Additives	Cyclamates	21,000
	Hexamethylenetetramine	65
	Nitrates	65,000
	Nitrites	3,500
	Saccharine	15,000
Pesticides	Dithiocarbonates	29
	Organophosphates	11
Cooking	Pyrolized amino acids	100
	Malondialdehyde	1
	Nitrosamine	2
AIR:		
Metals and other pollutants	Lead	6
	Asbestos	0.01 - 0.1
	Benzene	100
	Formaldehyde	100
	Halogenated hydrocarbons	20
ALCOHOL:	Acetaldehyde	10 grams
TOBACCO:	Tar	50,000
	Acetaldehyde	2,000
	Formaldehyde	50
	Benzene	200
	N-nitroso compounds	2
	Polycyclic hydrocarbons	1
	Cadmium	1 - 4

The data presented in table 7.4 can distinguish the effects of occupation. For example, for the general population, exposure to a toxic chemical (e.g., styrene, tetrachlorethylene, trichlorethane) is low, perhaps under 5 µg/day. However, the person who works around this chemical may have an exposure of 0.1 to 3 g/day. Variations in geographic distribution, diet, occupation, etc. can greatly alter the individual exposure.

Using studies such as this, it was proposed and hotly debated that chemical pollution induces about 80 new cases of cancer per day [5, 110, 216]. This data does not account for other chemical pollution-induced diseases, or the indirect carcinogenic action of chemical pollutants through the activation of oncogenes. Oncogenes are altered versions of the normal genes that control cell growth. Great controversy as to the exact impact of chemicals on people and the environment exists among geographic regions, even within the same country.

7.3 ACTIVATION OF POTENTIAL CHEMICAL CARCINOGENS

The development of a malignancy is a long-term, complex process involving multiple steps. It involves the accumulation of genetic or cellular lesions that finally translate into a tumor.

Environmental factors are believed to greatly influence the first steps of carcinogenesis, which may be reversible. Precarcinogens, or promoters, are compounds with weak carcinogenic properties that have metabolites with strong carcinogenic activity. Some promoters, such as phorbol esters, act by perturbing the surface of cellular membranes. Geneticists say that chemical pollutants and ionizing radiation are able to activate preoncogenes or to inactivate normal genes whose expression is necessary to maintain a nontransformed phenotype. Some preoncogenes activated by radiation or chemical pollution are identical to viral oncogenes or their variants. It has been demonstrated that malignant cellular transformation induced by chemical pollution or radiation occurs with the expression of viral genes and the methylation of genes that express globulins. Patients with xeroderma pigmentosum show a sensitivity toward exposure to chemical pollutants due to deficiencies in DNA repair enzymes [173, 231, 241].

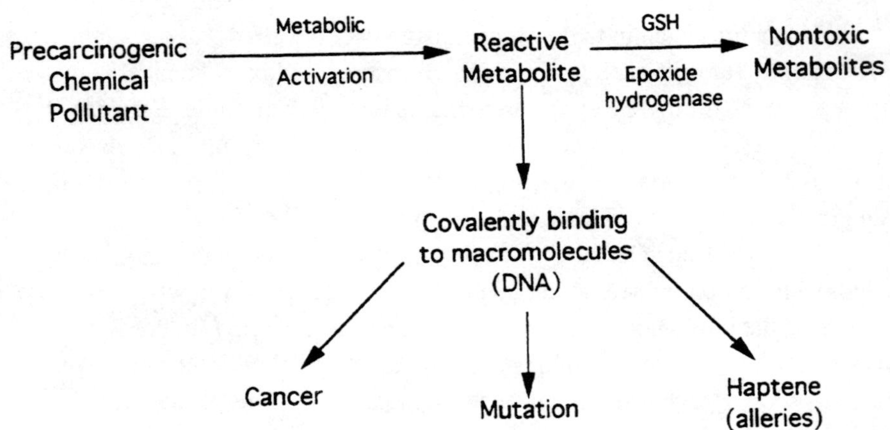

Figure 7.1. Proposed steps in the modification of gene expression.

These comments reflect the complexity of the tumor activation process. The first steps involve biotransformation reactions of chemical pollutants or the action of ionizing radiation. Long-term, chronic exposure to promoters leads to the accumulation of toxic metabolites. This leads to significant, durable modifications of DNA. It is accepted that DNA is the principle target for carcinogens. However, some specific cell surface receptors are also involved in the expression of modified gene expression. A summary of these events is given in figure 7.1.

The first step in biotransformation involves the formation of metabolites by the activity of monooxygenase. These metabolites are not stable and are, commonly, detoxified by the second phase reactions, involving the nonspecific action of glutathione transferase or the more specific action of epoxide hydrolase. These reactions, being enzymatic, possess a reversible nature, while the covalent binding of metabolites to DNA is much less reversible. The covalent binding of metabolites to DNA is required for the events that lead to cancer, mutation, or other diseases (hepatotoxicity, aplastic anemia, teratogenesis, immunopathology, drug-induced lupus). However, the initial work at studying these steps in carcinogenesis focused on metabolite activation.

The classic work on activation was done by studying polyaromatic hydrocarbons (PAH), part of which is presented in figure 7.2. The metabolism of PAH principally takes place in the endoplasmic reticulum and is catalyzed by the monooxygenase enzyme system. These enzymes are present in most

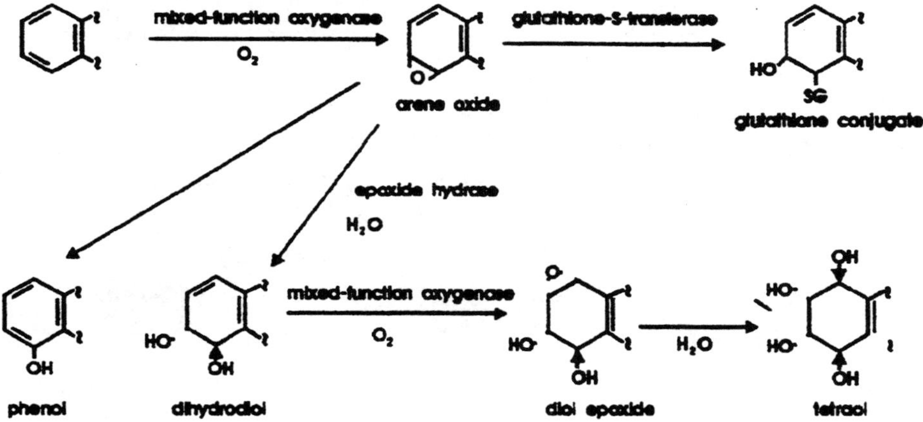

Figure 7.2. The polyaromatic hydrocarbons (PAHs) BP, DBA. DMBA, and 3-MC are potent carcinogens. BP and DBA are common environmental contaminants found in most of the populated areas of the world. The bay region, identified in BP, is a region between two adjacent, fused rings. The K region, a feature common tomany of the more potent carcinogenic PAHs, consists of an unsubstituted carbon-carbon bond flanked by two aromtic rings. Its chemical reactivity lies between that of a fully aromatic bond and localized olefinic bonds.

Figure 7.3. The principle metabolic pathway of polyaromatic hydrocarbons (PAHs) involves initial oxidation to an arene oxide, catalysed by the mixed-function oxygenases. The compounds formed are shown in stereochemical format with a black wedge-shaped line indicating a single bond projecting above the plain of the molecule and a dashed line indicating a bond below that plain. Oxidation of the PAH may occur in more than one molecular region. Quinones and further oxidation product not shown may also be formed.

mammalian tissues and serve the important function of detoxifying xenobiotics. These biotransformation reactions produce a mixture of isomeric phenols, dihydrodiols, quinones and their glucuronic and mercapturic conjugates, depending on the initial xenobiotic. Arene oxides and epoxide intermediates are hydrated to dihydrodiols by epoxide hydrolase or have glutathione added by glutathione transferase.

Arene oxides and epoxides are metabolites that are involved in covalent bonding to DNA. Pure preparations of these compounds show exceptional mutagenic and tumoragenic activity. The diol epoxides of PAH exhibit greater chemical reactivity than the K region of arene oxides. The formation of these compounds is given schematically in figure 7.3.

Current research with other classes of carcinogenic chemicals indicates the majority is procarcinogens that only become carcinogens through metabolic activity [49, 54]. James Miller and Elisabeth Miller [186], University of Wisconsin, stated the hypothesis that the ultimate, active, form of chemical carcinogens is a strong electrophile. These are molecules that contain electron-rich sites, such as amino groups, that readily form covalent bonds. Diol epoxides act like reactive electrophiles because of the ease with which the epoxide ring opens, generating an ionic intermediate with an electron-deficient (i.e., positively charged) carbon atom adjacent to the aromatic ring. The charge is stabilized by delocalization over the aromatic π bonds. Other classes of carcinogens, including nitrosamides, nitrosamines, and acetylaminofluorene, are potential alkylating agents, which can be converted directly to electrophilic species in animal cells. It should be stressed that not all mutagens are carcinogens. For example, acridine orange is highly mutagenic, but exhibits no detectable carcinogenic activity. It is also claimed that approximately 90% of chemical carcinogens exhibit mutagenicity under screening tests such as the Ames Test. However, among the 10% that are the exceptions are such well-known carcinogens as DDT and other chlorinated compounds.

The IARC considers polyaromatic hydrocarbons (PAHs) to be potential carcinogens, and their wide distribution in the environment is indicated in table 7.4. Most PAHs are formed from the incomplete combustion of fossil fuel (especially coal). The most active PAH contains four to six condensed rings. All PAH are highly lipid soluble and possess a great affinity to bind covalently to proteins and nucleic acids.

The covalent binding of PAH to DNA primarily takes place at the amine on the number 2 carbon of guanosine with the formation of a pair of adducts with the epoxidic form of PAH. Diol epoxides (Fig. 7.3) are formed during the enzymatic oxygenation of the olefinic bond of PAH. Other common covalent binding sites of active metabolites to DNA (nucleophilic centers) include the nitrogens at position 7 in guanine, and at positions 1 and 3 in adenine. In proteins the affected locations are the nitrogen at positions 1 and 3 in histadine, the sulfur in cysteine, and the terminal amine in valine.

The lengthy period from initiation of chronic chemical exposure to the accumulation of sufficient DNA lesions to be clinically significant suggests the intensity of the competition between the activation of xenobiotics and defense mechanisms. Numerous models of the way in which gene expression is finally modified have been proposed. Studies on these reactions are complicated by the multiple pathways available to biotransformation. For example, the PAH benzo-a-pyrene may be metabolized in the liver into over 40 metabolic products. One-third of these products possesses an epoxide structure. The product with the greatest affinity for binding with DNA is anti-benzo-a-pyrene-7, 8-diol-9, 10-epoxide [54, 208, 272].

Most variations in biotransformation end products result from enzyme diversity. Aryl hydrocarbon hydrolase (AHH) from liver (involved in PAH metabolism) has 76 molecular variants and epoxide hydrolase has 63 molecular variants. These polymorphic enzymes differ in their inducibility [1, 10, 97].

Organisms possess repair enzymes that enable them to survive the continuous oxidative stress and other causes of DNA damage [6, 44, 138]. These enzymes also possess polymorphism that may be related to differences in regulatory genes, structural genes, or post-transcriptional modification on enzyme function. A variety of human syndromes that predispose individuals towards cancer, "failure to thrive," and premature aging have been associated with known or suspected deficiencies in the repair of damaged DNA. Examples include xeroderma pigmentosum, Fanconi's anemia, ataxia telangeictasis, retinoblastoma, etc.

The repair of radiation- and chemical-induced base substitution in DNA has been observed for exposure to UV radiation, N-acetoxyaminofluorene, and some PAHs [49, 138, 293]. Various exogenous environmental factors (physical, chemical, and biological) alter the accuracy and rate of DNA repair. In 1972, Sato and Yamamoto showed that cells infected with certain viruses have altered post- replication repair of DNA.

Most research has been focused on the possibility of measuring the DNA-metabolite adducts in people chronically exposed to chemical pollutants [110, 122, 130]. Much of this effort has been aimed towards using DNA adduct technology to detect and identify environmental carcinogens through analysis of the adducts formed by reactions of genotoxic carcinogens with DNA or proteins. The quantitative determination of protein adducts in the hemoglobin of circulating erythrocytes, for example, provides a sensitive method to monitor human exposure to genotoxic chemicals and may provide a basis for estimating the dose of toxic metabolites that ultimately reach cellular DNA in tissues. As various authors have shown, gas chromatography-mass spectrometry and immunochemical methods are capable of detecting these products. A rapid method, based on the determination of adducts to the N-terminal valine residue of hemoglobin is suitable for routine monitoring of occupational exposure to simple epoxides and their precursor alkenes.

In following this line of research, Wright [298] and Fernandez [79] demonstrated a linear correlation between ethylene oxide exposure (10 - 30 ppm) and alkylation of DNA in various tissues. In people chronically exposed to toxic chemicals, S-methylcystein, phenols, and thioglycolic acid were found to be elevated by 10 - 50 fold. These compounds are the final metabolic products of methychloride, benzene, and vinylchloride, respectively. Only chronic exposure to levels greater than 1 - 5 ppm produce significant effects. Some noncarcinogenic compounds, such as organic solvents (xylene or toluene) produce increased levels of urinary thio ethers. But, as Van Doorn [71] showed, the excretion of urinary thio ethers is strongly correlated to cigarette smoking. Several methods developed during the past 10 years are now sufficiently sensitive and noninvasive to be used for the monitoring of a variety of carcinogenic-DNA adducts in humans [60, 88, 284].

As previously mentioned, the measurement of biochemical indicators are considered preferable to direct DNA analysis for determining the degree of exposure to chemical pollutants. This is because biochemical indicators are more easily available and samples can be obtained in larger amounts, sensitive analytical methods are available, and the half-life for excretion is known. Tornquest [284] has demonstrated the proportionality between DNA adduct formation and hemoglobin adduct formation, making it simpler to measure hemoglobin adduct to determine exposure.

Reactions of electrophiles with nucleophilic sites result in the formation of an adduct:

$$\text{RX} \quad + \quad \text{Y}^- \quad \xrightarrow{K} \quad \text{RY} \quad + \quad \text{X}^-$$

electrophilic compound	nucleophile	reaction product (adduct)	remaining group

where K is the rate constant for the reaction.

Therefore, the dose of a genotoxic compound can be determined by:

$$D = \int C dt = \int_t [RX] dt$$

where, if the concentration (C) of a metabolic intermediate (RX) is given in millimolar (mM), the dose is expressed in mM/hour.

Like other proteins, hemoglobin contains a number of nucleophylic sites (Y) that are reactive towards genotoxic compounds. For simple alkylating compounds N-terminal valine, cysteine-S, histidine-N, and methionine-S are the major reaction sites on hemoglobin.

The determination of hemoglobin adducts following exposure to aromatic carcinogens includes a number of steps, some of them quite difficult: isolation, derivatization with fluorinating agents, and analysis by gas chromatography-mass spectrometry [284].

In human exposures, which are mostly chronic, steady levels of adducts can be measured.

$$[RY]_{ss} = a/K \text{ (where } a = K \cdot Dh)$$

$$\text{and } [RY] = K\int_t [RX] dt = K \cdot D$$

$$\text{so } D = 1/K \cdot [RY]$$

Dh = the average dose rate per hour, K = the rate of disappearance of the adduct, and ter = the life-span of the erythrocyte. Therefore, the average dose per hour in blood having a steady state level of a hemoglobin adduct is:

$$Dh = (1/k) \cdot [RY]_{SS} \cdot (2/ter)$$

In DNA, the constant K varies from 10^{-2} to 10^{-1}. This corresponds to a half-life for the adduct of 0.5 to 5 days. In contrast, protein adducts are not subject to repair. In this case, a steady state level is reached after about one erythrocyte life-span (ter), which is about 4 months. Therefore, the rate constant for adduct formation is:

$$K^- = 2/ter = 6.6 \times 10^{-4}/hour$$

This method is of value for both exogenous exposure to contaminants (such as air pollutants and food components) and endogenous production of reactive intermediates in normal metabolism.

7.4 MUTATION

The transmission of the genetic code through replication and transcription, as well as the presence of repair processes, ensures a correct copying of the nucleotide sequence to a new molecule. Modification of the nucleotide sequence of the DNA molecule is not exclusively associated with xenobiotic compounds. It is believed that the mutation rate is one mutation per 100 million cellular divisions. Each mutation is assumed to introduce one error into the genetic code. In humans, approximately 10^{12} cell divisions occur per day, suggesting as many as 1 million mutations can occur each day. However, there are several types of possible mutations:

a) mutations at the gene level, altering the gene's code and, therefore, the function of a gene

b) mutations at the chromosome level, caused by rearrangements of the order of genes on a chromosome

c) mutations occurring in anaphase, leading to an altered separation of chromosomes between the daughter cells (polyploidy). This may occur when a transverse scission instead of the normal transverse scission occurs.

d)   latent mutations, where the genetic defect is not expressed in each
     generation, but only when two modified genes associate into a new
     allele.

The existence of such a high rate of mutation does not necessarily imply a
high rate of carcinogenesis. Mutation/promoter studies indicate cells can have
their DNA undergo mutation, but remain "quiescent" for a long period of
time. If these mutated, quiescent cells are then exposed to any of a range of
promoter chemicals or physical agents, the cells become active. It seems that
these chemical and physical promoters alter the cell membrane fluidity,
setting off a series of biochemical reactions. These reactions range from
altered membrane transport to derepression of certain genes. Multiple models
of mechanisms leading to mutation have been proposed (Fig. 7.4).

The DNA macromolecule contains regions of high risk where the
probability of mutations is high. These centers are places where activated
metabolites easily bind and have a nucleophilic character due to the presence
of the N-7 and O-6 of guanine and the N-1 and N-3 of adenine.

Mutagens can be categorized by their structure or physical properties.
Thus, ionizing radiation produces free radicals that react with nucleotide
bases to form compounds like 5,6-dihydrothymidine, which has a distinctive
ESR signal [190, 166, 192]. Anti-tumor drugs like the purine analogs (6-thio-
guanine) or pyrimidine analogs (6-amino-uracyl) inhibit the biosynthesis of
RNA. The same effect occurs with alkylating agents, such as yperite, and
epoxides, peroxides, ethylenediamines, and diazo compounds.

As research methods improved, new sources of mutagens in the
environment were discovered, such as those formed during food preparation.
Most of these mutagens are a result of cooking meat and fat-containing food,
which contain tiny concentrations of polyaromatic hydrocarbons (1 ppb), at
temperatures of more than 150° C. The compound produced under these
conditions has been identified as 2-amino-3-methylimidazo-4, 5-quinoxaline,
and has mutagenic activity towards *S. typhimurium*. This compound is
produce during Maillard browning reactions, so the formation is inhibited by
sodium bisulfite. Stich [272] has demonstrated that the concentration of this
compound is 2 - 15 times lower in caned meat than grilled meat. Microwave
cooking seems to substantially reduce the production of mutagens during
cooking.

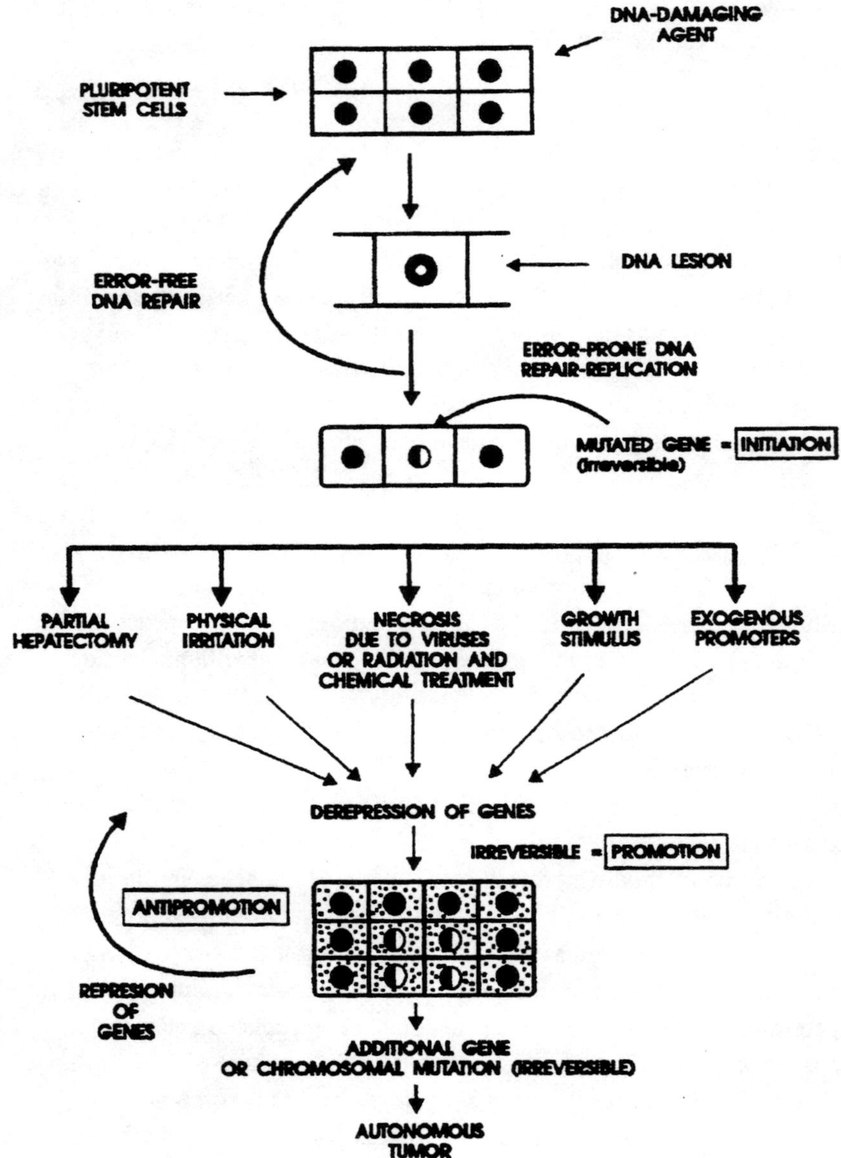

Figure 7.4. Multiple modes by which initiated cells can be promoted. Promotion is viewed as a process allowing cells with a specific carcinogen-induced mutation to multiply, enabling them to reach either a "critical mass" in order to become resistant to the anti-proliferative influences of normal cells or to increase the chance that one of these mutated cells will receive a second mutation.

Kasai and Nishikimi [137], of the Japanese Research Center for Cancer, observed that glucose crystals, following heating at 200° C for 20 minutes, become mutagenic towards bacteria such as Salmonella. The compound formed was identified as methylglyoxal. They also identified other mutagens in food: 8-hydroxyguanosine and 8-hydroxydeoxyguanosine. The Japanese team discovered that the hydroxylation of deoxyguanosine, a step required to produce mutagenicity, may occur at 37° C by:

- other mutagens such as polyphenols and aminophenols
- clastogenic compounds such as catechols and hydroquinone
- carcinogens such as 3-hydroxyquinoline.

The reaction requires the presence of oxygen. Reactive oxygen species are formed by the Fenton reaction, leading to an increased yield of hydroxylation. The same reaction is produced by ionizing radiation. Similar hydroxylation reactions are produced by heavy metals and asbestos. Considerable experimental proves that mutagen producing hydroxylation proceeds through the intermediate formation of OH$^\bullet$. Additional proof of this pathway is its inhibition by ethanol and dimethylsulfoxide. Similar modifications were obtained following the treatment of calf DNA with catechol-H_2O_2-iron, asbestos, hydrogenperoxide, or X-rays.

Another group of Japanese scientists [200, 210, 302] examined common drinks for the presence of mutagens. As shown in table 7.5, these scientists found variable amounts of mutagens, mostly in the form of dicarbonyls (glyoxal and its ethylated and methylated derivatives). The mutagenicity of dicarbonyls has been tested using *S. typhimurium*. The most potent are methyl glyoxal (toxic at 100 μg/ml), glyoxal (toxic at 20 μg/ml), and methyl glyoxal (toxic at 25 μg/ml).

The formation of mutagens during cooking or commercial processing of foods was also studied by Krone and Jeh [150]. According to their studies, the greatest amounts of mutagen formation occur when butter-coated and breaded convenience foods are heated in vegetable oil at about 200° C prior to packaging and freezing. Potent mutagens have been found to be present in beef extracts, with 2-amino-3-methyl imidazo-quinoline or quinoxaline contributing nearly 100% of the mutagenic activity.

Table 7.5
Amount of dicarbonyls in various beverages and foods. After Nagao [198].

	Glyoxal (μg/ml)	Methylglyoxal (μg/ml)	Ethylglyoxal (μg/ml)
Bourbon	0.39	1.5	0.42
Apple brandy	0.33	0.32	0.43
Wine	0.97	0.57	0.92
Instant coffee	0.34	1.6	0.70
Brewed coffee	0.87	7.00	1.90
Black tea	0.02	0.05	0.10
Green tea	trace	trace	0.34
Soft drink	-	1.40	-
Bread	0.30	0.79	0.40
Soy sauce	4.90	8.70	8.40
Soy bean paste	4.20	5.10	4.20

By drinking three cups of Turkish coffee a day, almost 1 mg/day of methylglyoxal is ingested. In spite of the fact that purified methylglyoxal is carcinogenic when injected into rats, no noticeable carcinogenesis occurs from drinking coffee. It seems that the combination of methylglyoxal and hydrogen peroxide exhibits both mutagenic and carcinogenic activity. Methylglyoxal ($HCO(CO)CH_3$) is a good electron acceptor, strongly inhibits free SH groups, and inhibits cellular division. Szent Gyorgyi [278] has reported methyl glyoxal is a natural regulator of mitosis. Nagao [198] showed the mutagenicity of coffee may be greatly enhanced or reduced by hydrogen peroxide.

The studies mentioned above raise the problem of hydrogen peroxide in organisms. This peroxide is continuously formed in organisms in the appreciable amounts of 1 - 10 mM/kg wet weight of tissue per hour [22, 218, 239]. Studies in cellular immunology demonstrate hydrogen peroxide is continuously formed by polymorphonuclear leukocytes and macrophages as a universal mechanism for killing pathogens [49, 137, 280]. Other agents, such as the cytostatic antibiotics adryamycin, bleomycin, and mitomycin, exert their cytolytic effect through the production of hydrogen peroxide. Hydrogen peroxide also exerts other biological effects:

a)   *in vitro*; damages DNA of viral or bacteria by the production of strand scissions (about 600 scissions per molecule at a concentration of 6 mM)

b)   significantly increases the mutation rate when catalase is inhibited by azide

c)   greatly increases the mutation rate in insects treated with stable peroxide prior to irradiation. Formaldehyde, a known mutagen, exerts its activity through the formation of hydrogen peroxide.

d)   hydrogen peroxide has a direct effect in forming tumor ascites cells.

e)   some precarcinogens, antineoplastic drugs and quinones exert *in vivo* cytoxic effects only following activation, during which $O_2^\bullet$ and H_2O_2 are formed

f)   hydrogen peroxide is a significant product of the *in vivo* action of UV and ionizing radiation

The direct action of hydrogen peroxide on nucleic acids is efficiently inhibited in aerobic organisms by the enzymes catalase and glutathione peroxidase.

Croton oil, obtained from the seeds of *Croton tiglium*, was once used in cosmetics but is now banned because it was found to be toxic to stem cells and to damage chromosomes. Methotrexate, a cytostatic drug, acts in a similar manner to croton oil by blocking pyrimidine synthesis at dihydrofolate reductase and promoting aberrant cell division. Croton oil also acts as a promoter in carcinogenesis. The active substance has been isolated from the oil and is found to be 12-o-tetradecanoyl-phobol-13-acetate (TPA). Croton oil (TPA) acts *in vivo* through the production of $O_2^\bullet$ and H_2O_2 and is inhibited by superoxide dismutase and catalase [44, 124].

Metabolism of paraquat, a herbicide, results in the activation of oxygen. Kidney cells exposed to paraquat become resistant to its effects through the induction of Mn-SOD. One feature of cancer cells is the low or absent level of Mn-SOD [26]. According to Oberley [208], during the activation of precarcinogens, a parallel activation of oxygen takes place. When the ability to synthesize SOD is lost, this activation of oxygen leads to DNA damage causing mutations and aberrant amplification of genes. The activation of genes associated with cell division may be a defensive reaction of the cells against mutagenesis. By increasing the rate of cellular division, the chance of some cells survival is increased. This increased rate of proliferation has been demonstrated in several tumor cell lines where the growth rate is 30 - 60 times that of normal cells [152, 165, 189, 224]. These oncogenes are found in extrachromosomal elements.

Given the potential frequency of spontaneous mutation and the exposure of organisms to carcinogenic and mutagenic compounds, the function of the defensive systems of cells is of great interest. Most of these systems focus on the repair of DNA lesions. These systems may be divided as:

a) Repair pathways where the lesion is removed without replacement of the phosphodiester link or

b) Repair pathways where the lesion is removed and the phosphodiester link is restored, frequently with the synthesis of a new DNA molecule. Repair of DNA lesions involves:

- a photoactivated enzyme that reverses pyrimidine dimers formed from exposure to UV light.
- insertion enzymes that replace missing purines
- alkyltransferases that are able to demethylate O-methylguanine, formed by the action of methylating or alkylating drugs (methyl nitrosourea).

c) Excision repair of damaged bases requires the presence of specific DNA glycosylases. These form apurinic and apyrimic centers that are subsequently recognized and removed by endo- and exonucleases, then filled by DNA polymerase and polynucleotide ligase. During repair by excision, the damaged structure is removed as an oligonucleotide. This mechanism has been demonstrated in the repair of damage from UV exposure and aflatoxin.

d) Repair by replication or unprogrammed synthesis of DNA ensures a correct *de novo* synthesis following UV exposure or treatment with cytostatic drugs. Theoretically this mechanism should act without errors, but in patients with congenital diseases, an increased frequency of mutations has been observed due to deficiencies in the repair mechanisms. A nonrepaired lesion can block correct replication of DNA.

Table 7.6
Products from DNA lesions eliminated in the urine. After Ames [5] and Cross [53].

Compound	nmoles/kg body wt.	molecules/cell
Thymine glycol	0.39	270
Thymidine glycol	0.10	70
S-Hydroxy methyluracyl	0.90	620
S-Hydroxy methyl-2'-deoxyuridine	trace	trace

Following the action of repair enzymes, deoxynucleosides are eliminated in the urine. Dr. C. E. Cross of the University of California, Davis devised a technique for measuring the products of DNA lesion in humans. The deoxynucleosides can be quantitated using high performance liquid chromatography (HPLC) and tested for mutagenicity using the Ames test.

Table 7.6 shows that under normal conditions 100 nmoles of DNA damage products are produced daily. Several scientists have shown that these products result from DNA damage and not from diet or intestinal flora [5, 200, 257, 269]. The amount of excreted damage product is equivalent to 10^3 molecules of thymine oxidized/day for each of the 6×10^{13} cells in the body. Since the products presented in table 7.6 are only four possible products of DNA damage, the total number of oxidative lesions formed in DNA/cell/day for humans is greater than 10^3.

7.5 INHIBITORS OF CHEMICAL CARCINOGENESIS

As might be expected, the DNA repair enzyme systems are not sufficient to prevent or reverse all of the effects of exposure to environmental carcinogenic compounds. Much research has been performed to discover the mechanisms and compounds protective against carcinogenesis. Most of these protective compounds act during the first step of the activation of promoters by inhibiting the metabolic activation or by enhancing the detoxification systems. Unfortunately, are only effective protective agents when administered prior to or concurrent with the carcinogenic exposure.

A. *Disulfiram* and its derivatives (diethyldithiocarbomate), when added to the diet, is an effective inhibitor of bladder or colon cancer induced by 1,2-dimethylhydrazine [54, 271]. The active metabolite of disulfiram is carbon disulfide (CS_2), which inhibits the oxidation of carcinogens. Surprisingly, some insecticides, such as ethylene bis-dithiocarbomate, contain a portion of their structure that is metabolized to CS_2. As Stich [127] has shown, disulfiram effectively inhibits the carcinogenic activity of diethyl- and dimethylnitrosoamine on the esophagus.

B. *Phenolic antioxidants* and ethoxyquin include several compounds used in the manufacture of canned food, such as butylated hydroxyanisole (BHA) and butylated hydroxytoluene (BHT). These natural and synthetic antioxidants are effective inhibitors of several known cancer promoters (Table 7.7). The greatest protective effect is obtained when the antioxidant is fed a week prior to the carcinogen exposure. In addition to ethoxyquin (a nonphenolic antioxidant), some phenolic antioxidants extracted from plants (*o*- and *p*-hydroxycynamic acids) have been tested for their protective effects. These studies have attracted much attention among scientists because the phenolic antioxidants are wide spread in natural foods. Once the protective action of these antioxidants was established, research was directed to locate the metabolic site of this carcinogenic inhibition.

A comparative study of the action of antioxidants on enzymes involved in biotransformation is presented in table 7.8. It should be remembered that these enzymes are primarily localized in the endoplasmic reticulum of liver and other organs. It is apparent from table 7.8 that an induction of epoxide hydratase takes place. This is a surprising result, because this enzyme is involved in the metabolic activation of PAH by producing diol endoperoxide. A simultaneous administration of a phenolic antioxidant and methyl cholantrene, a potent carcinogen, produced an increase in the activity of epoxide hydratase.

Table 7.7

Inhibition of carcinogenesis by phenols and ethyloxyquin (EQ). After Wattenberg [290].

Carcinogen	Tumor location	Antioxidant
Benz-a-pyrene	Lung, mouse	BHA, EQ
7,12-Dimethyl-benz-a-anthracene	Lung, mouse	BHA
	Skin, mouse	BHA, EQ
Dibenz (a,b) anthracene	Lung, mouse	BHA
Diethylnitrosamine	Lung, mouse	BHA, EQ
4-Nitroquinoline-N-oxide	Lung, mouse	BHA, BHT, EQ
Uracil mustard	Lung, mouse	BHA, EQ
4-Dimethylaminoazo-benzene	Liver, rat	BHA
Azoxymethane	Large intestine	BHA

Table 7.8

Influence of dietary antioxidants on biotransformation enzymes in rat liver.[a] After Kahl [134].

MEOS component	Control	EQ	BHT	BHA	PB
Epoxide hydratase	8.8±0.3	36.9±0.8	25±2.2	14.2	15.5±2
Aryl hydrocarbon hydroxylase	0.75±0.1	0.5±0.03	0.61±0.05	0.60	0.560.07
Ethoxy coumarin hydroxylase	1.7±0.1	2.3±0.1	3.2±0.4	1.82	4.90.3
Cytochrome P450	0.94±0.1	1.12±0.1	1.07±0.01	0.97	1.80.2
Cytochrome b5	0.61±0.05	1.0±0.06	0.93±0.06	0.80	0.73±0.05

[a]Units of enzyme activity / mg cell protein.

The great variation in inductive effects of antioxidants emphasizes the conclusions of Poland and Kende [223] regarding the metabolic activation of the extremely toxic chemical, 2,3,7,8-tetrachlorodibenzo-p-dioxin (TCDD), also known as dioxin. They demonstrated that during its metabolism, dioxin covalently binds to a cytoplasmic receptor. This complex penetrates the cellular nucleus triggering the induction of epoxide hydratase and glutathione transferase. This activity, which has also been demonstrated for steroid hormones, is actually a biochemical defense system. Another procarcinogen, polyaromatic hydrocarbons, competes for the complexing site with dioxin and the activity of antioxidants is probably by the same action.

C. *Flavonoids* are another group of compounds found in plants. Of these, 5,6-benzoflavone, quercitin, and rutin (pentahydroxyflavone) have been found to be potent inhibitors of pulmonary adenoma in rats [189, 192, 198, 243]. However, their effectiveness is less than phenolic antioxidants. These flavones also induce aryl hydrocarbon hydroxylase to a similar extent as polyaromatic hydrocarbons. Quercitin is mutagenic *in vitro* but harmless *in vivo*. The metabolism of flavonoids produces oxygen activation, but their slow absorption in the digestive tract reduces their biological properties [148, 243]. The flavonoids are not equal in their capacity to inhibit the formation of pulmonary adenoma and ranks β-napthoflavone > tangeretin > quercitin > nobiletin > rutin.

D. *Indoles*, such as indole-3-carbinol, 3,3'-diindolyl-methane, and indole-3-acetonitrile, occur in edible cruciferous vegetables. The cruciferous vegetables include Brussels sprouts, cabbage, cauliflower, and broccoli and are widely consumed. When present in the diet, all three indoles inhibit benzopyrene-induced neoplasia of the forestomach and pulmonary adenoma. All three indoles induce the activity of biotransformation enzymes towards benzopyrene, phenacetin, and hexobarbital. In experiments where neoplasms occurred at sites distant from the site of administration of the carcinogen, such as mammary tumor formation resulting from the oral administration of benzopyrene, the inhibitory effectiveness paralleled the ability of the indole to induce aryl hydrocarbon hydroxylase activity.

In one study, diets containing large amounts of cabbage and Brussels sprouts were fed to normal human volunteers aged 21 to 32 years. The results indicated the diets rich in vegetables enhanced the metabolism of antipyrene and phenacetin. Another study [243] showed an inverse correlation between the magnitude of cabbage consumption and the occurrence of colon cancer. Other investigations have been published in which an inverse relationship has been found between the consumption of vegetables such as lettuce, celery, and tomatoes and cancer of the stomach or precancerous lesions in the stomach [2, 185].

E. *Retinoids*, a group of compounds related structurally to vitamin A, possess a strong capacity for cellular differentiation and proliferation. In rats with vitamin A deficiency, no differentiation of stem cells into epithelial cells occurs and an excessive accumulation of keratin takes place. Only recently has the importance of retinoids in cellular differentiation and proliferation come to be appreciated [3, 148, 302]:

- Animals with avitaminosis A who are exposed to chemical carcinogens exhibit a marked sensitivity compared to control animals treated under similar conditions [28, 43, 232].
- Vitamin A and synthetic derivatives administered in the preneoplastic stage inhibits the production of epithelial tumors [185].
- In cell culture, retinoids inhibit tumor development [161, 196].
- When added to cell culture after tumor development, retinoids may strongly influence the subsequent cell proliferation and differentiation.
- Retinoids regulate the development of the vascular system in chicken or rat embryos.
- Using recombinant DNA methods, it has been proven that retinoids regulate the development and differentiation of mesenchimal precursor cells [26, 166, 288].
- In human promyelocytic leukemia, retinoids induce the terminal differentiation of malignant leukocytes into mature granulocytes.

It is known that β-carotene is the essential precursor to retinol (vitamin A). Since carotenes occur naturally in fruits and vegetables, it accounts, in part, for the protective effect against cancer of a vegetable diet. Carotenes, in addition to being precursors for vitamin A, are also excellent antioxidants and free radical scavengers, especially for peroxyl and hydroxyl radicals.

In a study published in 1991 in the *American Journal of Clinical Nutrition*, Alfthan, et al. found a strong protective relationship between serum β-carotene levels and lung cancer, bladder cancer, and melanoma, but no protection against rectal cancer [2]. However, recent scientific discussions have questioned the protective action of carotenes.

F. *Vitamin C* (ascorbic acid) indirectly participates in a number of key redox enzymatic reactions. A very specific role of ascorbic acid is to act as a trap for nitrite, particularly at a pH below 6.0. Nitrate-induced stomach cancer is prevalent in the orient and western Latin America and esophageal cancer is common in China. Ascorbic acid is thought to partly account for the reduced risk of these cancers in populations where fruits (oranges, kiwi, etc.) and vegetables are widely eaten. Vitamin C has also lowered the incidence of UV-induced skin cancer in mice.

Exposure to toxic pollutants, including tobacco smoke, increases vitamin C requirements, partly because of its increased metabolism under these circumstances. The treatment of cancer with megadoses of vitamin C, as suggested by Linus Pauling 30 years ago, is again under debate. It has also been noted that AIDS patients have a lowered serum vitamin C level. Depending on environmental conditions, particularly the nitrate levels of food and water, ingestion of greater amounts of vitamin C than called for by the US RDA may be beneficial, especially if obtained through the increased ingestion of fruits and vegetables.

G. *Tocopherol*, or vitamin E, is an excellent nitrite trapping agent, especially when coupled with vitamin C. Experiments have shown that the toxic effects of ozone on the lung is largely prevented by vitamin E. This vitamin protects the circulating red blood cells and other essential cell systems against reactive oxygen species. Reduced risk for coronary heart disease has been noted with increased ingestion of vitamin E, which may in part be due to its red blood cell protective effect. Vitamin E has been observed to reduce the cardiotoxicity of anthracycline cytostatic drugs.

The nonspecific protective mechanisms of the vitamins against cancer are thought to be:

1. Prevents formation of active carcinogens.
2. Increases effectiveness of detoxification.
3. Control of transformed cell replication.
4. Control of malignancy expression.
5. Control of the differentiation process.
6. Control of cell-cell communication.

It is reasonable to conclude that the nonspecific protective defenses that act against free radicals are also act efficiently against free radicals formed during the metabolic activation of carcinogens.

7.6 ESTROGEN-LIKE COMPOUNDS

Recently a new aspect of chemical pollution has entered the debate. This is the presence of estrogen-like compounds and their biological effects. These pollutants do not affect the body through free radical formation, but are mentioned here because of their significance. This is a world-wide problem with frightening prospects, so a historical overview is appropriate.

First came DDT, the famous polychlorinated insecticide that destroyed mosquitoes, but, due to its unusual stability, remained unaltered in the soil and fats of vertebrates for decades. DDT (trichlorethane) is lipophilic and can produce numerous effects including death, hepatoma, or brain damage. Due to its chemical stability and wide use, DDT can be found everywhere in the world, but it is found in higher concentrations in some places. One such place is Lake Apopka, Florida. Here, Dr. Louis Guillette discovered sterility in all male alligators in the lake, as well as a decreased sperm count in the men of the area. A decrease in sperm count by as much as 50% has been reported in many or most countries. Many aromatic compounds have structural similarities with the natural female sex hormone estrogen (Fig. 7.5). A list of these compounds would be very large and include many recognized chemical pollutants: polychlorinated biphenyls (PCB), insecticides (DDT), atrazine and endosulfan (chlorinated compounds used to bleach paper), diethylstilbestrol (a synthetic estrogen), polycarbonate plastics (bis phenol A, nonyphenol), and especially aromatic hydrocarbons (PAH such as 3,9-dihydrobenzantracene).

As a hormone, estrogen acts within the body at specific receptors localized on cell membranes. Estrogen-like compounds can interfere with or compete for this binding, altering biochemical processes, especially in the reproductive system. The result can be sexual development in both males and females gone seriously askew.

An even more dramatic consequence has recently been reported by Davis and Bradlow [61]. In the blood of women with breast cancer, the level of DDT or a metabolite of DDT was found to be increased. It seems that estradiol and estrogen-like compounds can be metabolized to a "good" product, 2-hydroxyestrone, or a "bad" product, $16\text{-}\alpha\text{-}$hydroxyestrone, which exhibits carcinogenic properties. Estrogen-like compounds seem to shift the metabolism towards the production of $16\text{-}\alpha$ -hydroxyestrone. Extensive physical exercise significantly increases the production of 2-hydroxyestrone.

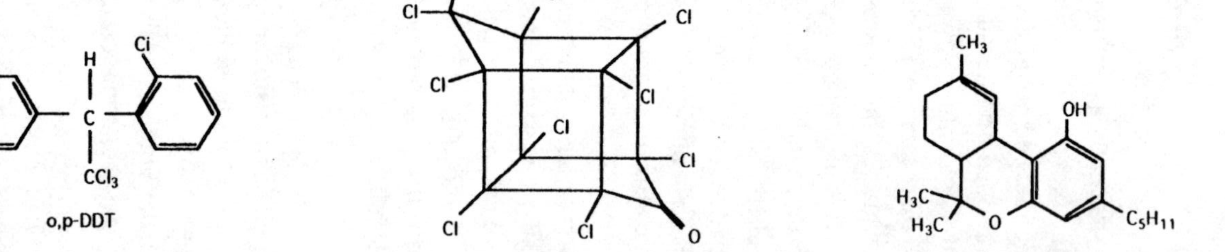

Figure 7.5. Chemicals with estrogen like properties.

In spite of a high exposure to estrogen-like compounds, breast cancer is low in Asian countries. The explanation may be a diet rich in certain vegetables, such as soy products, cabbage, and broccoli, which are rich in antioxidants. These vegetables also contain phytoestrogen, which is structurally related to 2-hydroxyestrone and seems to have beneficial effects on humans. Many edible plant products, such as French beans (string beans) and pomegranates contain some phytoestrogens that might combat cancer. This might help explain why women in Japan have low rates of breast cancer when they eat tofu, soy sauce, and miso; all rich sources of estrogens. Complex compounds in "healthy" foods like broccoli might nudge estrogen metabolism towards the cancer fighting form. These studies are very new and much more research is needed.

7.7 CELL PROLIFERATION AND APOPTOSIS

Some of the most intense research efforts in cell and molecular biology are focused on determining the mechanisms that regulate mammalian cell proliferation. The control of cell proliferation involves a variety of genetic and biochemical factors that remain poorly understood. A cell is able to divide according to a well regulated genetic program, but in response to some irritating factor or harmful condition, the proliferation process can be modified. So, depending on the nature and magnitude of the irritating factor, the cell may release a defensive compound, or proliferate. But, if these defensive actions are overwhelmed, the cell shifts to necrosis. Apoptosis is a form of programmed cell death. It is characterized by cell shrinkage and nuclear fragmentation. It differs from necrosis, which is characterized by cell swelling, in that apoptosis is an active process that may be triggered by a number of internal or external stimuli. Therefore, apoptosis may serve in many cell types as a "failsafe" device to prevent cells from proliferating uncontrollably when a persistent stimulus is present.

The dead cells often fragment into apoptotic bodies that are rapidly phagocytized and digested by macrophages. In this way, dead cells are rapidly removed without spilling their contents. In contrast, during necrosis, a pathological form of cell death that results from overwhelming cell injury, cells swell and lyse, releasing cytoplasmic material that often triggers an inflammatory response. Consequently, apoptosis is of central importance in the maintenance of homeostasis in living organisms.

Table 7.9
Modifiers of apoptosis

Inhibitors -   Physiologic: growth factors, amino acids, sex hormones   Pathologic: viruses, tumor promoters, protease inhibitors
Inducers -   Physiologic: TNF*, neurotransmitters, calcium, glucocorticoids   Pathologic: reactive oxygen species, bacterial toxins, radiation, chemicals,   viruses, ethanol, cytostatic drugs

*TNF = tumor necrosis factor

Not surprisingly, the initiation of apoptosis is carefully regulated (Table 7.9). Many different signals that may originate within or without the cell have been shown to influence the fate of cells. These signals include a great variety of compounds. Chemical pollutants may induce apoptosis directly or by intermediate reactive oxygen species. As suggested by McCord and Sarafian, [180, 250], superoxide (and other reactive oxygen species) act as a triggering signal for apoptosis and hyperproliferation. A condition of increased oxidative stress, as occurs in the inflammatory process, may serve as an ideal signal for surviving cells to proliferate and replace the lost cells, and for fibroblasts to migrate into the area and repair the wound. The involvement of apoptosis as a response to exposure to a chemical pollutant has been demonstrated for liver cancer [93, 250].

CHAPTER 8

RESPIRATORY PROBLEMS

8.1 OVERVIEW

Chemical pollutants reach the lungs in two ways: directly by inhalation and indirectly by blood circulation. Once in the lungs, chemical pollutants interfere with functional processes, particularly oxygen-carbon dioxide exchange. Chemical pollutants also produce biochemical lesions, leading to characteristic diseases such as pneumonia, bronchial asthma, pulmonary edema, and adenocarcinoma. Exposure to some pollutants, such as specific organic solvents, fumes, or mineral dust, have acute clinical effects in addition to the pathologies that develop from chronic exposure (asbestosis, silicosis). Pulmonary edema can also develop as a side effect of treatment (hyperbaric treatment or intensive care) or because of aerospace accidents. Acute or chronic pulmonary lesions may also result from indirect exposure through blood circulation. Treatment with cytostatic drugs (bleomycin, cyclophosphamide), some antibiotics, or antihypertensive drugs can produce lung failure. These side effects must be weighed by health professionals in making treatment decisions [71, 80, 221].

The World Health Organization (WHO) estimates that nearly 625 million people around the world are exposed to unhealthy levels of sulfur dioxide, and that more than a billion people, one in five of the world's population, are exposed to excessive levels of particular pollutants. In recent years, despite a

Table 8.1
Selected chemical pollutants in the atmosphere

Source	Chemical
Climate and wind	Minerals, plant, and animal particulates; gases: CO_2, H_2S, NH_3; spores; pollen
Volcanic eruptions	Particulates; gases (SO_2); ash
Industry	Coal dust; ash; gases: CO_2, H_2S, NO_2, CO_x; fumes; Ng; As
Cement factories	Mineral dust
Chemical factories	Gases: Cl_2, SO_2, SO_3, H_2S; solvents; hydrocarbons; phenols
Exhaust gases	CO; NO_x; Pb; hydrocarbons; solid particulates
Home furnaces	Solid particulates; ash; soot; CO; SO_2
Smoke	CO; NO_x; Cd; Pb; hydrocarbons

decrease in pollutants from high sulfur fuels, studies in various countries have shown an association between acid aerosols and morbidity and mortality, especially among people with asthma. The health effects of airborne particles, soot and solids, are difficult to separate from the effect of other pollutants, with which they often interact.

As seen in table 8.1, the sources of airborne pollutants is impressively varied and widespread. In addition, there is great potential for the superposition of several components, especially particulates and gases like SO_2, CO, and NO_x. The data varies with geographic area and specific country. In the USA, automobile exhaust gases account for 43% of all airborne emissions, while in Russia it amounts to only 15%.

Airborne pollution is often the first to be noticed because of its smell and sometimes its visual effects. Airborne pollution is a chronic problem that may, under certain conditions, become acute or even deadly. For example, many large cities have a chronic problem with automobile exhaust that becomes an acute problem of smog when exhaust chemicals interact with fog and sunlight. London, England possesses a sad history of periods of smog-induced mortality that occurred in 1873, 1880, 1948, and 1952. In December 1952, a dense smog covered London killing about 4,000 people. Other industrial cities have had similar but less severe episodes: Glasgow (1909 and 1925), Meuse Valley (1930), Donora, Pennsylvania (1948). The concentration of sulfur dioxide in these events reached a peak of nearly 4,000 $\mu g/m^3$, more than 10 times the maximum level established by WHO as safe to breath for one hour. The victims, mostly young children and people over 65, died of heart and lung problems, including bronchial irritation, bronchospasm, dyspnea (difficulty breathing), and cyanosis. A summary of these effects is presented in table 8.2.

As seen in tables 8.1 and 8.2, sulfur dioxide and particulates, such as soot, are the main airborne products from burning fossil fuels. Given the correct meteorological conditions, cold moist air and the presence of certain metal catalysts in polluted droplets of water, they combine to form aerosols of sulfuric acid. This easily reaches the lungs during breathing and is probably the component of winter pollution that most damages the lungs. The depth to which particles penetrate the respiratory tract depends on the particle size (Fig. 8.4). Particles smaller than 2 μm penetrate more deeply. Breathing through the mouth also contributes to increased penetration of particulants since the nose normally filters out much of the larger particles [17, 59, 207].

- Hygroscopic particles increase in diameter when they absorb water, so are filtered out more readily than nonhygroscopic particles of the same dry size.
- The amount of particulate that deposits in the lungs increases as the breathing rate increases. Thus, the depth of penetration is increased by exercise because of the faster, deeper breathing.
- The smaller the particle, the more that reaches deep into lung tissues. Particles 10μm in diameter and larger are essentially completely removed in the nasal passages. Particles about 5 μm in diameter are nearly 100% removed by the nasal passages and upper respiratory tract. 25% of particles 0.5 μm in diameter are removed from the inhaled air. Thus, the efficiency of the upper respiratory removal of particles decreases with decreasing particle size and becomes negligible at about 1 μm diameter.

Diesel fumes have now replaced coal burning as the major source of particulate pollution in urban areas, amounting to one-third of total particulate emissions [255, 281]. Using occupational studies, the IARC has concluded that diesel engine exhaust is probably carcinogenic to man.

Table 8.2

World Health Organization standards for pollution and the effect when they are exceeded.

	Airborne particulates	Sulfur dioxide	Nitrogen oxides	Carbon monoxide	Ozone	Benzene
Source	Diesel (90% in towns), coal burning	Fossil fuels, power stations (73%), diesel exhaust	Motor vehicles (45%), power stations (35%)	Incomplete combustion of fossil fuels, tobacco smoke	Photochemical reactions between nitrogen oxides and hydrocarbons	Emission and evaporation from gasoline engines. Highest at filling stations and in cars
Health effects	carrys acidic gases and volatile hydrocarbons into lungs. May be carcinogenic.	Bronchitis, bronchiospasm (especially in asthmatics)	Respiratory irritation.	Reduces oxygen capacity of the blood. Headaches, impaired concentration, triggers angina, can trigger arhythmas and cardiac arrest. Can retard fetal growth.	Coughing, impaired lung function, eye, nose and throat irritation, headache. Aggravates asthma and bronchitis.	Causes leukemia.
Environmental effects	Soiling of buildings, reduced visibility, odor.	Main consituent of acid rain. Damages plants and aquatic life.	Responsible for one-third of the acidity of acid rain.	Oxidizes to carbon dioxide, contributing to the greenhouse effect.	Greenhouse gas. Damages crops, trees, plastics, rubber, and paints.	
WHO air quality standard:						No safe level, carcinogenic.
1 hour average		$350\,\mu g/m^3$	$400\,\mu g/m^3$	$30mg/m^3$	76 - 100 ppb	
8 hour average				$10mg/m^3$	50 - 60 ppb	
24 hour average	$120\,\mu g/m^3$	$125\,\mu g/m^3$	$150\,\mu g/m^3$			
1 year average		$50\,\mu g/m^3$				
Frequency of exceeding	Regularly.	1 hour average, regularly throughout Britian.	Busy roadside locations.	8 hour guideline exceeded 24 days at one London site in 1988.	Several times duringsummer, 1989. One site in Devon reached 135 ppb.	

Once chemical pollutants enter the respiratory tract, the consequences are varied and usually affect several levels of the respiratory tract.

- At the tracheal and bronchial level, for particles between 2 and 10 μm and for irritant gases, short-term exposure results in cough and increased mucus production.
- At the bronchial level, for particles between 0.01 and 2 μm and for irritant gases, a complex array of reactions occurs, including inflammation, metaplasia, and vasoconstriction.
- At the alveolar level, particles smaller than 0.5μm and gases like NO_x, ozone, sulfuric acid, and sulfur dioxide injure the tissue and reduce the ability to convey oxygen to the blood.

Generally, irritant gases, such as SO_2, NH_3, H_2S, formaldehyde, and acid vapors damage the upper respiratory tract. The more lipid soluble the gas, the easier penetration occurs. Depending on their structure and concentration, some gases can produce lethal effects through their inhibitory effects on oxygen transport. Carbon monoxide, nitrogen oxides, and hydrogen sulfide form stable complexes with hemoglobin, such as carboxy-methemoglobin or sulfurhemoglobin. These complexes are incapable of transporting oxygen.

Chronic exposure to atmospheric pollutants is the cause or aggravating factor for several diseases, including allergic rhinitis, rhinopharingitis, chronic bronchopneumonia, asthma, emphysema, and pulmonary cancer [63, 80, 221, 270]. In all these conditions, bronchial and pulmonary lesions can be observed, accompanied by cilial destruction and cellular metaplasia. As is the case for occupational pulmonary diseases, pathological modifications due to atmospheric pollution can be detected by radiological, clinical, and functional examinations. The wide variation in clinical manifestations indicates the precocious, reversible state is long, indicating the natural defense systems act efficiently for a lengthy period. According to WHO, chronic bronchial pneumonia is the most widespread respiratory affliction. This disease is defined as persistent cough lasting three consecutive months, extending over at least two consecutive years. But individual, meteorological, and geographical factors may strongly influence the clinical manifestations.

8.2 INFLUENCE OF METEROLOGICAL FACTORS: SMOG, OZONE, AND ACTIVATED OXYGEN

Unlike other means of exposure to pollution, the respiratory rout strongly depends on meteorological factors, including air currents and winds, temperature variations, humidity, ozone concentration, and photochemical reactions occurring in the atmosphere. The correct evaluation of all these factors is still very difficult and often impossible. The spreading of the radioactive cloud after the Chernobyl accident is a recent example of the implications of an accident that potentially could affect an entire continent. This accident clearly demonstrated the lack of knowledge we have concerning the effect of radioactive and chemical fallout on large areas, including the effects on the soil, plant and animal biological cycles, etc. [118].

8.2.1 METEROLOGICAL FACTORS

Unless there is some great event, like a major volcanic eruption, the air has a remarkably constant composition of 80% nitrogen ($3.8*10^{15}$ tons), 20% oxygen ($1.15*10^{15}$ tons), 0.05% carbon dioxide, traces of noble gases, and a varying amount of water vapor. The main layers of the atmosphere are presented in figure 8.1. For our discussion of pollution, the troposphere and stratosphere are of greatest interest, especially the ozone layer located at an altitude of 20 - 50 km. Unfortunately, man has succeeded in altering the ozone concentration in some locations. The question if there have been alterations in the oxygen and carbon dioxide content is still being debated.

The process of evaporation and condensation of water vapor takes place under solar heating, with an estimated 536 calories being required to evaporate one gram of water. This huge energy expenditure warms the atmosphere, producing the movement of masses of air with different temperatures. Normally, the lower layers are warmer than the higher layers. But, in calm weather, where vertical currents are absent, thermal inversion

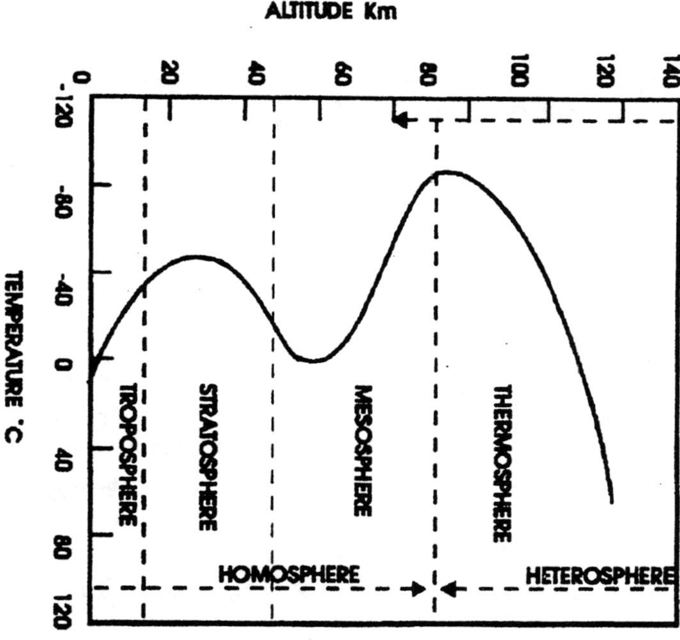

Figure 8.1. Vertical distribution of the average temperature of the atmosphere and atmospheric zones.

can occur with a zone of warm air being formed under a zone of cold air. An inversion slows the air circulation and can persist for days or weeks. During this time, high local concentrations of chemical pollutants are trapped close to the ground. It was such an inversion that resulted in thousands of deaths in the Meuse Valley, London, and Donora.

Recent studies on meteorological phenomena have revealed the circulation of dust, chemical pollutants, and radioactive materials in the stratosphere. The historic volcanic eruptions (Krakatoa, St. Helens) have contributed to our understanding of how widely airborne materials can spread.

Air humidity is another important factor in airborne pollution. Fog, for example, provides hydration to convert some gases to acids, and to trap solids.

Studies have attempted to predict the diffusion of a pollutant from a point source. Generally, the concentration of a pollutant in the soil is proportional to the strength of the polluting source and inversely proportional to the square of the height at which the pollutant is released.

This brief summary of meteorological factors involved in the spread of pollutants serves to demonstrate the complexity of the problem of predicting the dispersal pattern of chemical or radioactive airborne pollutants.

8.2.2 SMOG

Smog has become the symbol of atmospheric pollution. Smog is a loose combination of fumes, smoke, oxidants, and fog. The word "smog" was introduced at the beginning of the last century by the English physician Des Voen, who discovered its implications in causing some respiratory diseases. This combination of smoke and fog resulting in intense pollution generated public outcries even in seventeenth century London. Presently, the most smog ridden cities are Los Angeles, followed by Tokyo, London, Genoa, Baku, Moscow, etc. Studies on the formation of smog were complicated by the fact that smog formation varies according to local conditions and the exact pollutants present.

In London, the major toxic component is sulfur dioxide (SO_2) resulting from the burning of coal. This gas easily penetrates into the depths of the lungs, destroying cilia and alveoli.

In Los Angeles and Tokyo, the formation of smog involved a completely different mechanism. In these areas, smog appears as a whitish to brown mist producing strong irritation in the eyes. Dr. A. J. Hoagen Smith, who studied this type of smog, found that photochemical reactions are an important factor. Car exhaust provides 80% of the hydrocarbons found in the atmosphere, where it reached 2,500 tons daily in 1960. In the Tokyo area, the discharge of nitrogen oxides has drastically increased from 31,000 tons per year in 1960 to 173,000 tons per year in 1970. Hydrocarbon release raised from 80,000 tons per year to 231,800 tons per year during the same period. In addition, oxidizing compounds derived from photochemical reactions have greatly increased and cause people to complain of eye and respiratory irritation. The concentration of oxidants can exceed 0.15 ppm during most of the year in Los Angeles, while in Tokyo this generally occurs from April through October, a period called the photochemical season. These reactions proceed most efficiently with warmer temperatures. In Los Angeles the outdoor temperature remains around 15° C year around. In Tokyo these temperatures are only reached during the spring, summer, and fall.

Nitrogen has six oxides and is noted symbolically as NO_x. Most important are NO (monoxide), which reacts with hemoglobin to form methemoglobin, and NO_2 (dioxide), which is a strong, toxic oxidant. NO_2 irritates the eye's conjunctiva and the respiratory tract, where it is capable of producing a frequently deadly pulmonary edema. NO_2 directly produces bronchial and pulmonary lesions. The accepted safe short-term exposure (1 hour) is around 2 ppm, a level that has been measured in Los Angeles. Even in plants, a NO_2 concentration of 1 ppm inhibits growth and photosynthesis and causes changes in the intracellular inorganic ions (K, Ca, Mg).

In the upper atmosphere the following photochemical reactions can occur:

$$N_2O_3 \leftrightarrow NO_2 + NO \tag{1}$$

$$2\,N_2O_5 \leftrightarrow 4\,NO_2 + O_2 \tag{2}$$

$$N_2O_3 + H_2O_2 \leftrightarrow 2HNO_2 + O_2 \tag{3}$$

Nitrogen oxides, in the presence of light, oxidize olefinic hydrocarbons:

$$\text{Olefines} + NO_x - \text{light} \rightarrow CH_3\text{-COO-}NO_3 \tag{4}$$

Peroxyacetyl nitrate (PAN), once formed, is responsible for the withered, metallic color of vegetation, due to its strong oxidative capacity. PAN directly damages amino acids, proteins, and nucleotide coenzymes, but its strongest affinity is towards small molecular thiols like glutathione.

Aldehydes and ketones may also participate in photochemical reactions by forming free radicals.

$$CH_3\text{-CO-}CH_3 \xrightarrow{\text{light}} CH_3^\bullet + CH_3CO^\bullet \tag{5}$$

$$R\text{-CHO} \xrightarrow{\text{light}} R^\bullet + HCO^\bullet \tag{6}$$

These reactions only hint at the complexity of chemical and photochemical reactions involved in the formation of smog and its biological consequences.

8.2.3 OZONE AND ACTIVATED OXYGEN

Ozone (O_3) is a natural component of the upper atmosphere and is essential to all forms of life due to its ability to protect the surface of the earth from UV radiation. The ozone layer is found between 20 and 50 km in altitude at a location having an increased oxygen content. Because of its UV absorbing effect, life on earth depends on the presence of ozone in the atmosphere at a suitable concentration. The drop in ozone concentration observed over Antarctica, South America, and Northern Europe legitimately raises concern.

Even though the topic has been studied for at least 30 years, the mechanism of ozone formation is not completely known. In order to produce ozone, the oxygen-oxygen bond must be split, requiring an energy input of more than 120 kcal/mol. This reaction occurs in the upper atmosphere.

$$O + O_2 \xrightarrow{\text{light, M}} O_3 \tag{7}$$

where M is a photocatalyst, such as NO_x. But reaction (7) is opposed by a wide range of other photochemical reactions involving donors (D) and acceptors (A):

$$O_3 \xrightarrow{\text{light}} O\text{-}O(^1D) + O_2(^1A) \rightarrow O + O_2 \tag{8}$$

in which reactive oxygen species can be intermediate products:

$$O\text{-}O(^1D) + H_2O \rightarrow 2OH^\bullet \tag{9}$$

$$O(^1D) + CH_4 \rightarrow CH_3^\bullet + OH^\bullet$$

$$O\text{-}O(^1D) + N_2O \rightarrow 2NO$$

$$O(^1D) + CClF_2 \rightarrow ClO^\bullet + \text{other products} \qquad (10)$$

The concentration of ozone in the atmosphere is actually a dynamic result of all these reactions.

It has been established that ozone is only able to absorb UV radiation between 240 and 320 nm. Solar radiation below 240 nm is absorbed by a mixture of oxygen and ozone. Intensive studies [146, 191] have shown that UV radiation below 240 nm (the A fraction) are strongly absorbed by DNA and are directly responsible for the appearance of skin cancer. A decrease of only 10% in the ozone concentration in the atmosphere will allow the increase of UV radiation at 305 nm by 20% and at 287 nm by 500%.

The effect of chloroflurocarbons ($CFCl_3$ and CF_2Cl_2) on ozone, which, in the past, have been widely used in refrigeration, spray cans, and plastic foams, is shown in figure 8.2. This scheme was presented to the International Congress of Geophysics in Baltimore in 1987. The interactions of ozone and nitrogen oxides form cycles in the atmosphere that may be fed or interfered with by man's activities. It should also be pointed out that reaction (9) and figure 8.2 demonstrate the formation of reactive oxygen species in the atmosphere [15. 146, 191].

However, ozone has another aspect, completely opposed to the first, beneficial, aspect. As mentioned, ozone is involved in the formation of smog. The maximum naturally occurring concentration of ozone in air is 0.1 ppm. But for asthmatics, this level produces respiratory problems. Even at concentrations of 0.08 ppm, ozone can cause coughing, sore throat, and breathing difficulty. Such respiratory crises have been observed in aircraft personnel flying over polar regions at 13 - 15 km altitude, but only during the period from February to April. Animal experiments indicate the lethal dose of ozone is around 10,000 ppm, except for man for whom it is only about 20 ppm.

The toxicity of ozone strongly increases in the presence of other compounds in the atmosphere, such as nitrogen oxides, hydrocarbons, tobacco smoke, etc. This property has been observed at concentrations of 0.35 ppm and is explained by ozone's high reactivity. Ozone rapidly damages proteins and amino acids, such as cysteine, tryptophan, and tyrosine, affecting enzymes containing these amino acids. Not surprisingly, ozone is very toxic

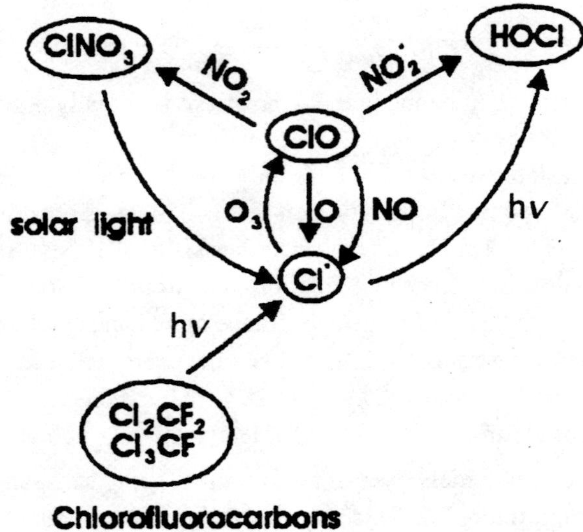

Figure 8.2. Photochemical reactions in the stratosphere. After Marx [178].

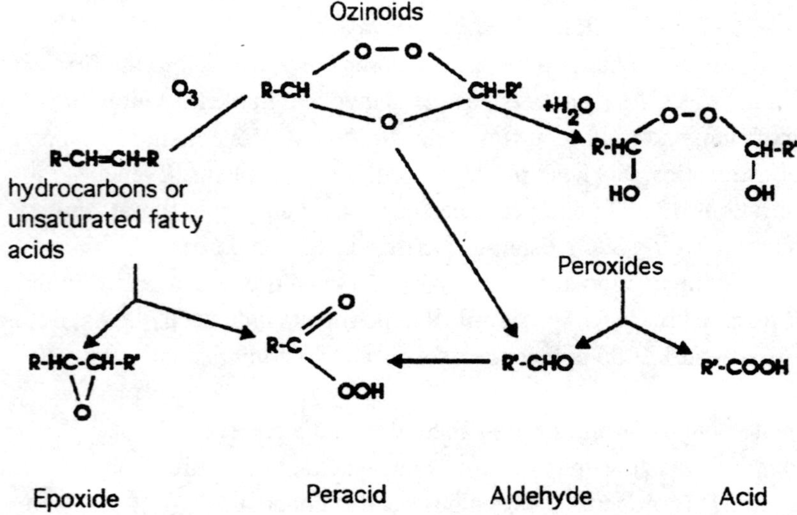

Figure 8.3. Ozone's reactions with hydrocarbons and unsaturated fatty acids. After Olinescu [211].

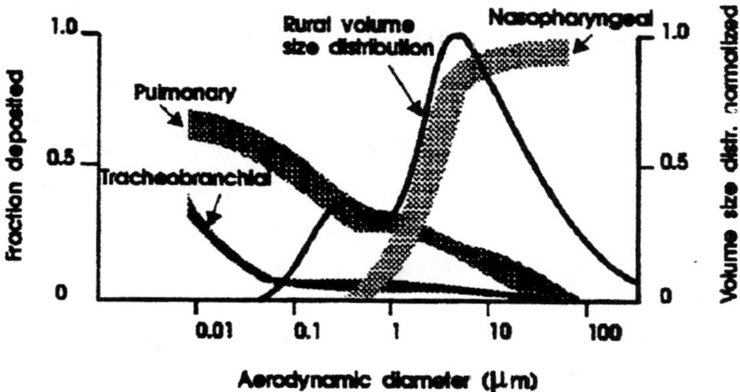

Figure 8.4. Collection efficiency (fraction deposited) of various parts of the human respiratory tract, and a normalized volume-size distribution of a rural aerosol. The ordinate is presented in a linear scale.

for children's lungs. At 0.6 ppm it bleaches textiles and can even damage rubber enough to cause car tires to blow out.

Toxic reactions of ozone are presented schematically in figure 8.3. As seen here, most of the toxic actions of ozone originate from its reaction with hydrocarbons or polyunsaturated fatty acids leading to the formation of epoxides, aldehydes, and other products having mutagenic activity through reactions with DNA [248]. Ozone and smog extracts act similarly, by producing inhibitory effects on macrophages and fetal pulmonary fibroblasts. Some of the toxic effects of ozone may be attributed to production of the OH$^{\bullet}$ radical [146]:

$$O_3 + H_2O \xrightarrow{\text{light}} O_2 + 2\,OH^{\bullet} \tag{11}$$

Sometimes the indoor safe concentration of ozone (100 μg/m^3) is exceeded in locations near air ionizers, movie projectors, electric motors, photocopy machines, and radar stations, where the concentration may exceed 200 μg/m^3 [146, 191].

8.3 RESPIRATORY DEFENSE SYSTEMS

Following respiratory exposure to a chemical pollutant, several defensive systems become activated, depending on the composition of the xenobiotic and the location along the airway. Adaptation can be observed to occur following chronic exposure to moderate concentrations of chemical pollutants. This adaptation starts at the molecular and cellular levels and then extends to the entire respiratory tract. A classic example is smokers and workers exposed to a constant environment of smoke and dust.

An adult requires about 6 liters of air per minute, taking up almost 250 ml of oxygen and eliminating 200 ml of carbon dioxide. Taking into account that during activity this figure can be multiplied as much as 20 fold, it is estimated that man uses approximately 20,000 liters of air in a day. This impressive figure illustrates the intensity of the exposure of the lungs to dust, bacteria, and chemical pollutants.

The structure of the respiratory tract includes vibrating cilia, mucus, surfactant, immunologic and biochemical systems.

Based on the nature of atmospheric pollutants, they may be divided into:

- Monotoxic agents: dust and smoke, which the organism treats as a foreign compound, activating mechanical and immunological systems.
- Irritative toxic agents: gases, aerosols, and some particulates, which produce physicochemical and immunologic reactions.
- Bacteria, spores, and parasites that stimulate the immune system.
- Allergens, which have a complex organic structure and induce an over-reaction from the body's immune system following repeated exposure.

Against this invasion of noxious agents present in inhaled air, the body reacts by activating local and general systems. Generally, an exposure to the pollutant induces first a physical reaction through the release of catecholamines from synaptic nerve ends (noradrenalin) or from the adrenal gland (adrenalin). Due to existing receptors in tissues, their action will produce vasoconstriction in the skin, lungs, and kidneys (α-adrenergic response) and vasodilation in brain capillaries (β-adrenergic response). Tachycardia, tachypenia, and hypertension can result from such an exposure, unless the dose is small or the person has experienced some adaptation [162].

Following exposure to a chemical pollutant, oxygen partial pressure decreases and carbon dioxide partial pressures increases. This causes precapillary sphincters to open. In addition, blood cells release a large range of biologically active compounds: prostaglandins, histamine, cytokines, serotonin. During adaptation to exposure to a chemical pollutant, local and general reactions occur through the nonspecific release of catecholamines. The catecholamine-induced vasoconstriction may be regarded as a defense reaction, but if the exposure is repeated chronically, may lead to hypoxia and acidosis.

Capillary constriction is compensated for by the modification of metabolic acidosis and alterations in the oxygen partial pressure. In some cases of exposure to toxic gases, vasoconstriction of pulmonary capillaries may lead to a nonspecific stasis. Such circulatory modifications, observed generally and in the respiratory tract are nonspecific, being also reported in some stress conditions and intense physical exertion.

8.3.1 REACTION TO DUST AND FIBERS

Man is constantly exposed to dust, particles, and microscopic fibers present in air. Presumably, during evolution, the respiratory tract became structured to avoid the intake of large amounts of dust, fiber, sand, etc. These particles are removed from the respiratory tract by the mechanical movement of hairs and vibrating cilia in conjunction with mucus and lung surfactant. This movement is rapid. The cilia vibrate at about 1,200 beats per minute, moving the mucus mass some 15 mm per minute.

The speed with which air moves through the respiratory tract varies markedly depending on the specific portion of the respiratory tract. The air flow rate is maximum in the trachea and in the turbulent zones of the bronchial tubes, decreasing in the alveolar spaces. This decrease in air speed permits the restraint and deposition of dust in the alveoli.

The pulmonary surfactant lining of the alveoli is composed of a lipid-protein complex, the major constituent of which is dipalmitoylphosphatidylcholine [17].

The restraint of dust by cilia and mucus and surfactant and its final removal by coughing and expectoration are components of a complex system of defense called clearance. Total clearance requires about 24 hours. Particle retained in the upper respiratory tract are rapidly evicted in 3 - 12 hours. Dust retained in the bronchial tubes are removed in 12 - 24 hours or more depending on the chemical composition and dimensions of the particle. The association of certain gases may reduce the rate of clearance, by reducing the activity of the cilia, so that weeks or months are required. Most of the dust smaller than 1 μm in diameter is permanently retained in the alveoli.

At the alveolar level another phenomenon related to protection occurs. This is phagocytosis by macrophages. Lung fibrosis related to the inhalation of dusts containing toxic particles, such as silica or asbestos, is an important occupational hazard. During an immunologic reaction against dust particles, reactive oxygen species are released by various inflammatory cells, with macrophages being most prominent. In addition, proteolytic enzymes and elastases are secreted. These act in concert with reactive oxygen species to denature proteins, destroy carbohydrates, and peroxidize lipids causing connective tissue alterations that lead to silicosis and fibrosis. Depending on the chemical composition, the amount inhaled, and the presence of toxic gases, inhaled materials can affect other organs, including blood and lymph nodes.

Asbestos has received particular attention in recent years for its cancer causing properties. Asbestos has been widely used in industry and construction and is still in use for some purposes: automobile brake pads and in some areas for water filtration. The main types of asbestos are crysotyle (88%) and crocidolite (10%), with both minerals possessing carcinogenic potential. The accepted limit for asbestos in air is less than 0.01 mg/m^3.

The lung possesses an elaborate defense system against reactive oxygen species. This system consists of several antioxidant enzymes (SOD, glutathione peroxidase, and catalase) and nonenzymatic antioxidants (glutathione, vitamin C, vitamin E, and carotenoids). These antioxidant systems are located in lung tissue, interstitial fluid, and circulating erythrocytes. A disturbance involving an excessive oxidative burst of phagocytes or a malfunction of the antioxidant systems can provoke damage and lesion formation in tissues. As seen in figure 8.5, polymorphonuclear leukocytes (PMNL) in workers with pneumoconiosis present higher chemiluminescence emission than those of healthy persons or to patients with pulmonary cancer.

It has been demonstrated in several studies [57, 255, 257] that the red blood cells of miners with pneumoconiosis have a significantly decreased content of glutathione. The erythrocytes of these miners are more sensitive to lipid-peroxidative damage induced *in vitro* by hydrogen peroxide. According to these studies, erythrocytes may act as a circulating antioxidant system. The decrease in glutathione content of the erythrocytes is a consequence of the long-term increased release of reactive oxygen species by the bronchial-alveolar cells. In addition to alveolar macrophages, neutrophils are also present in great numbers in inflamed areas. Neutrophils secrete substantial amounts of reactive oxygen species and cytokines (interleukin-1 and tumor necrosis factor).

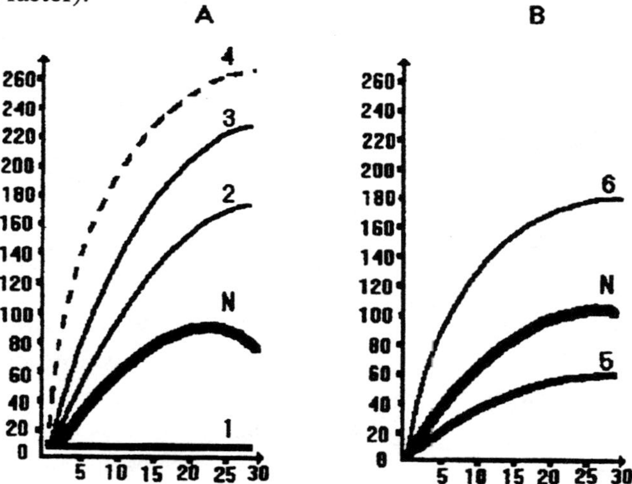

Figure 8.5. Chemiluminescence of PMN from workers with pneumoconioses (A) and from patients with lung cancer (B). The normal physiological values are noted with 'N'. For all groups, the control samples without opsonized zymosan are found in group 1. The other groups are different individuals. Immunosupression was observed only in patients with long cancer treated with cytostatics (line 5). After Olinescu [214].

demonstrated that in patients with adult respiratory distress syndrome (ARDS) the antioxidant enzymes may predict the development of the disease. ARDS is an acute inflammatory process characterized by neutrophil accumulation, edema in the lungs, and progressive hypoxia. ARDS occurs unpredictably as a complication in patients with sepsis. At the initial diagnosis of sepsis (6 - 24 hours before the development of ARDS) Mn-SOD and catalase are significantly increased in serum. The increase in these antioxidant enzymes occurs only in patients who will develop ARDS and reflects the activation of intrinsic antioxidants mediated by massive cytokine release by the neutrophils.

The involvement of reactive oxygen species released during phagocytotic activity of macrophages and PMNLs in the ethiology of pneumoconiosis is a result of an overactive defense mechanism. This mechanism was elucidated through a great many studies performed both *in vitro* and *in vivo* [96, 129, 268]. Gupta and Keew [102] demonstrated that during experimentally-induced pulmonary fibrosis in rats, great amounts of lipid peroxides are formed. These lipid peroxides form because of the action of reactive oxygen species on the lipid component of the pulmonary surfactant. The release of reactive oxygen species during elevated phagocytosis in the lungs stimulates an increase in collagen synthesis and conjunctive tissue, favoring the fibrotic process [114, 122, 221, 268].

Reactive oxygen species-induced formation of lipid peroxides was also observed in styrene toxicity [283]. A most interesting example is the biochemical consequences of diesel exhaust. As shown by Carraro [35] and Li [160], cars with diesel engines release more particulate exhaust than cars that use gasoline engines. Diesel exhaust produces dramatic decreases in lung and liver cell viability. Rats treated with components of diesel exhaust showed significant decreases in glutathione and free SH groups in blood and tissues. The cytotoxic action of extracts from diesel exhaust is reversed by thiol containing compounds, such as cysteine or N-acetylcysteine or by albumin and other proteins with a binding capacity. Since 1986, exposure to environmental tobacco smoke has been widely regarded in scientific circles as a cause of lung cancer. In that year, US and international health organizations concluded that passive smoking increased lung cancer risk by about 30%. In England, active smoking kills 20 times more people than traffic accidents and 40 times more than infectious diseases.

The certainty that tobacco smoke causes lung cancer is supported by the discovery of over 60 chemical carcinogens in tobacco smoke (table 8.1 lists five). Cigarette smoke contains several inducers of biotransformation enzymes, by which precarcinogenic compounds are converted into their active forms. The increased levels of chemiluminescence produced by PMNLs from smokers demonstrates the involvement of oxygen free radicals and their consequence.

The mechanism by which cigarette smoking is postulated to cause ischemic stroke includes increased fibrinogen and platelet aggregability, reduced cerebral blood flow (arterial vasoconstriction), and formation of atheroma [15, 116, 285].

The enormous resistance of people to the effects of active and passive smoke is connected to the biochemical adaptation of the antioxidant systems. Such adaptation can provide some protection, at least up to a certain point.

8.3.2 ASTHMA AND RESPIRATORY ALLERGIES

It is accepted that pollution increases the sensitivity of an organism to environmental allergens, inducing immunologic reactions that create a hypersensitive state. Thus, exposure to chemical pollutants and over active phagocytosis can be factors in respiratory allergies and bronchial asthma. The list of environmental pollutants involved in allergic phenomena consists of a wide range of chemicals: propellants from deodorants and cosmetic sprays, insecticides containing pyrethra or organophosphates, detergents, bleaching agents containing chlorine, drugs, etc. These chemicals may act alone or in combination with dust, pollen, and powders already present in the atmosphere.

Theoretically, organic and inorganic chemicals do not directly produce allergic reactions with cutaneous or respiratory manifestations. But, the following reaction is generally accepted and occurs in the presence or absence of light:

Chemical compound + Protein → Antigen (covalent complex)

Once formed and having entered the organism, the covalent complex between a protein and a chemical compound may act as an antigen through its interaction with mast cells and other immunologic systems. This contact-type of allergy has been demonstrated for photoallergies and bronchial reactions of

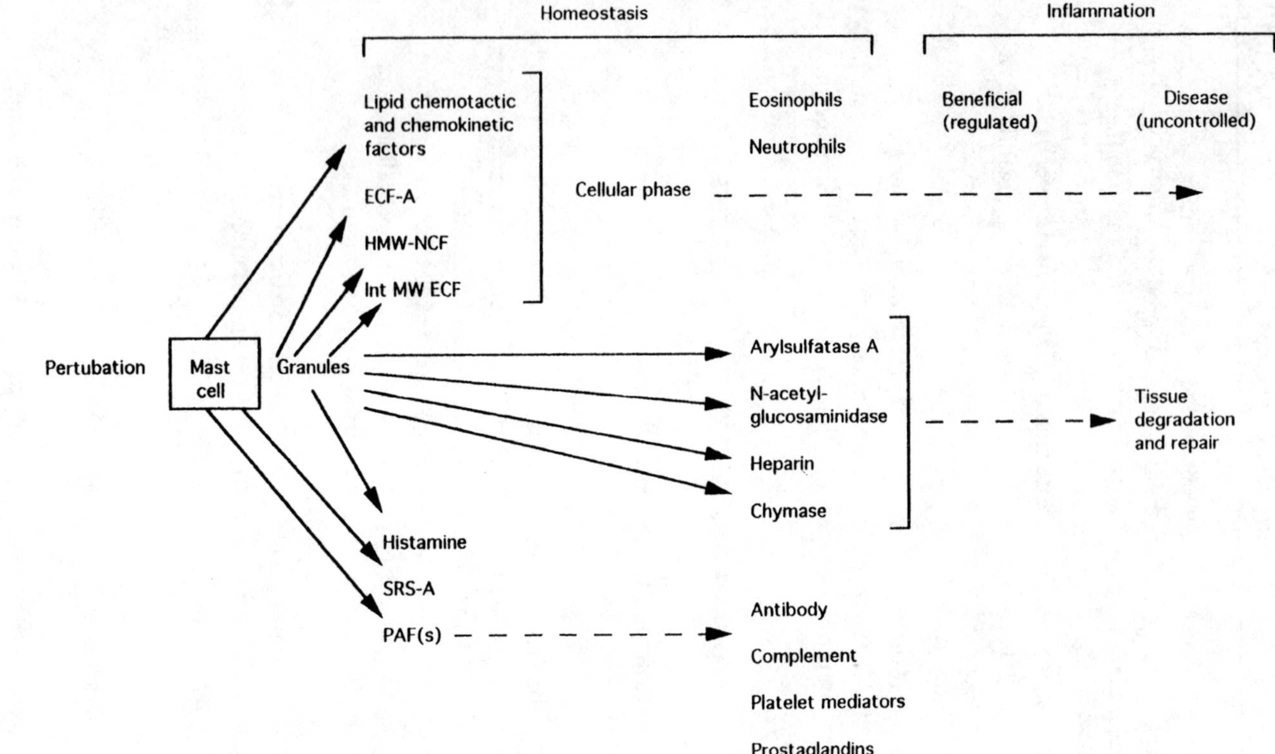

Figure 8.6. Schematic representation of the effects of mast cell activation on the host. The fully regulated effects of activation are considered to be homeostatic, whereas excessive activation is inflammatory, irrespective of whether or not the end result is beneficial or detrimental. Inflammatory, by definition, requires subsequent repair, whereas homeostasis refers to physiologic pertubations of the microenvironment at the cellular level. ECF- A = eosinophil chemotactic factor of anaphylaxis, NCF = neutrophil chemotacic factor, SRS-A = slow reacting substance of anaphylaxis, PAF = platelet activating factor.

immediate, late, or dual type. Allergy is simply an immunologic hypersensitivity with harmful tissue and physiologic effects. The biochemical mediators of allergic reactions of immediate hypersensitivity involving immunoglobulin E (IgE) are fairly well known (Fig. 8.6).

From our point of view, the activation of mast cells and eosinophils results in the release of a cascade of biologically active substances. These compounds include the slow-reacting substance of anaphylaxis (SRSA) (a leukotriene), prostaglandins, thromboxanes, and histamine and trigger many immunologic reactions. These are mostly unpredictable but have biological consequences such as bronchial spasm or cough. Even at this stage, the organism continues to resist by releasing bronchodilating substances like prostaglandin E_2 (PGE_2). But in bronchial asthma, the generation of substances with bronchial constrictive effects is greater than those of the opposite action [18].

The production of bronchial constrictors depends on the release of arachidonic acid from the membrane. This is used in the production of leukotrienes, prostaglandins, and thromboxanes with the intermediate production of reactive oxygen radicals. The release of these compounds and the resulting constriction of the bronchial tubes, etc. should be viewed as a defensive reaction by the organism to decrease the penetration of the xenobiotic compound.

Allergic reaction is triggered in an asthmatic crisis due to the imbalance between broncho-constricting compounds (SRSA) and those with the opposite action (PGE_2). This release of compounds synthesized from arachidonic acid (Fig. 8.7) is a dynamic process. Thus, 5,8,11,14-eicosatetraenoic acid inhibits the release of histamine in asthma patients, and in allergies with cutaneous symptoms. As seen in figure 8.7, both branches of the arachidonic acid cascade are regulated through enzymatic activity. Aspirin, indomethacin, phenylbutazone, and other nonsteroidal anti-inflammatory agents inhibit cyclooxygenase. But in some people, these agents can trigger an asthmatic attack due to an increased activity of lipoxygenase, favoring the release of leukotrienes having vasoconstrictor activity towards the upper respiratory tract [42, 162]. The increased lipoxygenase activity is due to the loss of substrate competition when cyclooygenase is inhibited, and is not a direct activation of lipoxygenase.

                    R. Olinescu, T. Smith and J. Hertoghe

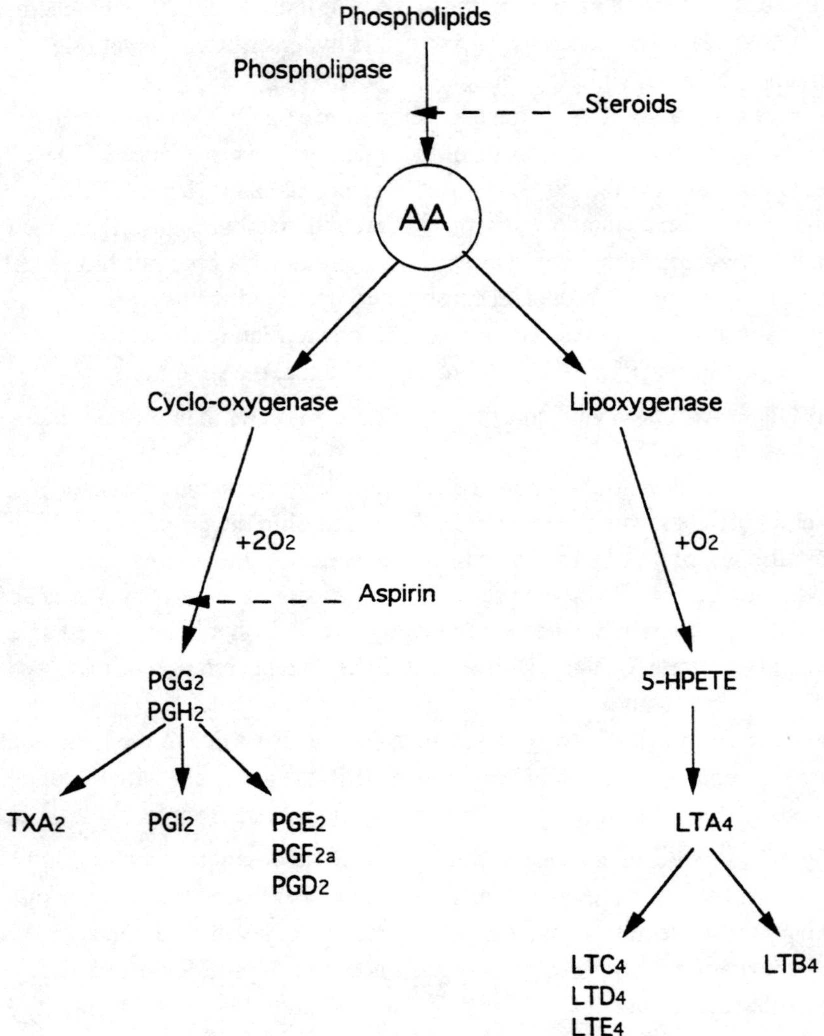

Figure 8.7. Cycloosygenase and 5-lipoxygenase pathways of arachidonic acid (AA) metabolism. PGG_2 and PGH_2 = cyclic endoperoxides; TXA_2 = thromboxane A_2; PGI_2 = prostacyclin; PGE_2, $PGF_{2\alpha}$, and PGD_2 = prostaglandins E_2, $F_{2\alpha}$, and D_2; 5-hydroperoxyeicosateraenoic acid; LTA_4, LTB_4, LTC_4, LTD_4, and LTE_4 =leudotrienes; - - - -> = inhibition.

As shown in figure 8.7, leukotriene A_4 (LTA_4) is converted into other, related compounds (LTD_4, LTE_4) by the activity of an enzyme similar to glutathione-S-transferase involving the addition of amino acids from glutathione. Consequently, the organism can control the formation of SRSA by the competition between the formation and decomposition of LTA_4 or the availability of glutathione [14, 94]. Modifications in the structure of glutathione transferase have been observed in hypersensitive persons.

Glutathione is also a coenzyme for another enzyme, glyoxalase. It has been shown that S-lactoyl-glutathione, the substrate for glyoxylase, inhibits the IgE-induced release of histamine through a modification of receptor affinity.

LTA_4 may also be converted to LTB_4, a leukotriene highly chemotactic towards macrophages and PMN leukocytes. LTB_4 contributes to the increased mobilization of these immunocompetent cells toward an infection site.

It must be remembered, of course, that during an allergic episode or asthma attack, mast cells (Fig. 8.6) play a key role by secreting large quantities of histamine, prostaglandin, leukotriene, thromboxane, lytic enzymes, kinins, etc., which amplify and diversify the biological response. Disodium cromoglycate, an early anti-asthmatic, prophylactic drug, inhibits the degranulation of mast cells and thus prevents mediator release. Ketotifen, a benzocycloheptathiophene, acts similarly, but is also active on basophil and neutrophil cells.

Eosinophils appear to be particularly important in the late-phase of the asthmatic inflammatory response, and are associated with bronchial hyper-reactivity. In addition to releasing superoxide radical, eosinophils can also produce several enzymes, including peroxidase. This enzyme, like myeloperoxidase from neutrophils, can use hydrogen peroxide and an appropriate halide as substrates to form hypochloric acid. This compound may contribute to tissue damage.

When cells are unnecessarily stimulated during an inflammatory response, the excessive production of oxygen free radical does not appear to serve any useful purpose. On the contrary, it contributes to the perpetuation of inflammation.

Table 8.3
The action of some gases on the lungs.

Symptom	CO_2	NH_3	NO_x	SO_2	Cl_2	O_2	O_3
Irritation	+ +	+ +	+ +	+ + +	+ + +	+	+ + + +
Dyspnea	+ +	+ +	+	+ + +	+ +	+ +	+ +
Bronchitis		+	+ +	+ + +	+ +	+	+ +
Pulmonary edema		+	+ + +	+ + +	+ + +	+ +	+ + +
Emphysema			+	+ +	+ +	+	+ +
Oxidation		+	+ +	+ + +	+ + +	+ + +	+ + +

8.3.3 THE ACTION OF SOME GASES

As stated earlier, the damaging effect of polluting gases on the organism starts at the molecular level due to their ease of penetration. The effect on the organism may be specific (carbon monoxide poisoning) or, more commonly, general.

The first observation to be made is that chronic intoxication with any toxic gas produces effects that are quite similar. However, the irritating gases (SO_2, Cl_2, NO_x, and O_3) are of the greatest importance (table 8.3). The second observation is that the production of pulmonary edema is a common manifestation of several gases and is closely related to other properties of the gas. The succession of clinical symptoms (cough, bronchitis, emphysema, and edema) is graduated and is a function of the dose, properties, and resistance of the individual [262, 267].

The involvement of oxygen and ozone in the pathology of respiratory diseases has only recently been recognized and greatly helps our understanding of these processes.

Pulmonary edema is the consequence of events leading to liquid accumulation in the interstitial spaces, followed by the alveolar spaces. The movement of fluids across microcapillaries can be described by Starling's relation:

$$J_{v,c} = K_{f,c} (\Delta P_c - \delta \Delta \pi_c)$$

where:

$J_{v,c}$ is the volume of liquid moving across the capillary wall,

$K_{f,c}$ is the hydraulic conductance of capillaries and equals 0.2 ml/min,

$\Delta\pi_c$ is the colloido-osmotic pressure gradient,

ΔP_c is the gradient of hydrostatic pressure acting on the capillary wall.

Because the lungs have a huge surface, making for a huge exchange capacity, the value of $K_{f,c}$ / 100g of tissue is about 0.2 ml/min.

Pulmonary edema can be the result of increased endothelial pressure against plasma proteins. When vascular pressure is increased, $\delta\Delta\pi_c$ will also increase across the capillary wall. This provides a buffering capacity to protect the capillary wall from increased pressure. This defensive system is the safety factor of pulmonary edema. When the capillary permeability toward plasma proteins increases, this factor is surpassed, the interstitial spaces are flooded, and edema occurs.

Table 8.4

Variations in fluid volume of pulmonary edema and elastase activity following the activation of leukocytes. After Smith and Thorpe [266].

Treatment	Edema µl plasma/ml fluid	Elastase u/ml fluid
Saline	7.6±2.8	2.3±0.8
Leucinphenylalanine (200µg/kg)	17.4±7.9	7.0±2.4
Phorbol myristate (20 µg/kg)	43.0±2.5	15.7±9.6
Phorbol myristate (40 µg/kg)	55.0±8.3	17.9±8.5
Glucosoxidase + glucose	112.8±47	17.2±4.7

According to Taylor [277], more than half of the factors and agents implicated in the formation of pulmonary edema (microemboli, hyperoxia, peroxides, SRSA, PG, etc.) are directly related to the activation of neutrophils and other phagocytic cells. Leukocytes are directly involved in damage to capillary endothelium. SOD incorporated into liposomes prevents the formation of edema.

It should be mentioned that other agents implicated in the occurrence of edema, such as paraquat, gases, hallucinogens, and napthylthiourea, are able to directly produce activated oxygen.

In 1987, Taylor used the technique of broncho-alveolar lavage to prove that great amounts of elastase could be recovered in the lavage fluid. Elastase is a protease directly involved in the damage of tissue. Normally it is inactivated by a natural inhibitor from the plasma. Following activation of leukocytes, reactive oxygen species are released, inactivating the inhibitor, allowing elastase to act freely. Table 8.4 lists the amount of fluid production and elastase that results in the lung when exposed to certain chemicals. Leucinphenylalanine is a chemotactic peptide, phorbol myristate is a strong stimulator of phagocytosis, and glucoso-oxidase + glucose is an enzyme system that releases hydrogen peroxide and superoxide.

During the formation of pulmonary edema, a great decrease in glutathione and free SH groups is known to occur. Another proof of the involvement of reactive oxygen species in the pathology of pulmonary edema came from studies using cobra venom factor (CVF), an extract that spectacularly increases capillary permeability and subsequently leads to pulmonary edema. The action of CVF is reversed by inhibitors of oxygen free radicals, such as catalase, dimethylsulfoxide, and desferrioxamine.

Therefore, there is strong evidence that oxygen activation is directly involved in the formation of pulmonary edema. Thus pulmonary edema is explained as one of the negative consequences of oxygen activation following exposure to chronic or acute pollutants or other toxic agents.

Now that the implication of leukocytes and phagocytosis in pulmonary edema is explained, we can present the sequence of steps occurring with exposure to gases and dust from the environment (Fig. 8.8). According to this scheme, the critical step is located at the point of phagocytosis. A moderate exposure to chemical pollutants induces an adaptation or resistance based on the control of oxygen free radicals, the recovery of epithelial cells, and restoration of a normal structural-functional relationship. But an acute or chronic exposure induces fibrosis due to an exaggerated release of reactive oxygen species and the accumulation of the resulting damage.

Most alveolar cells are type I, meaning they are sensitive to the toxic action of free radicals or active metabolites and do not divide because of their topographic differentiation. Their recovery following a moderate toxic exposure takes place by proliferation of alveolar epithelium cells (type II). Interstitial cells and those of capillary endothelium may divide.

Another efficient line of defense is the enzymatic antioxidant systems that

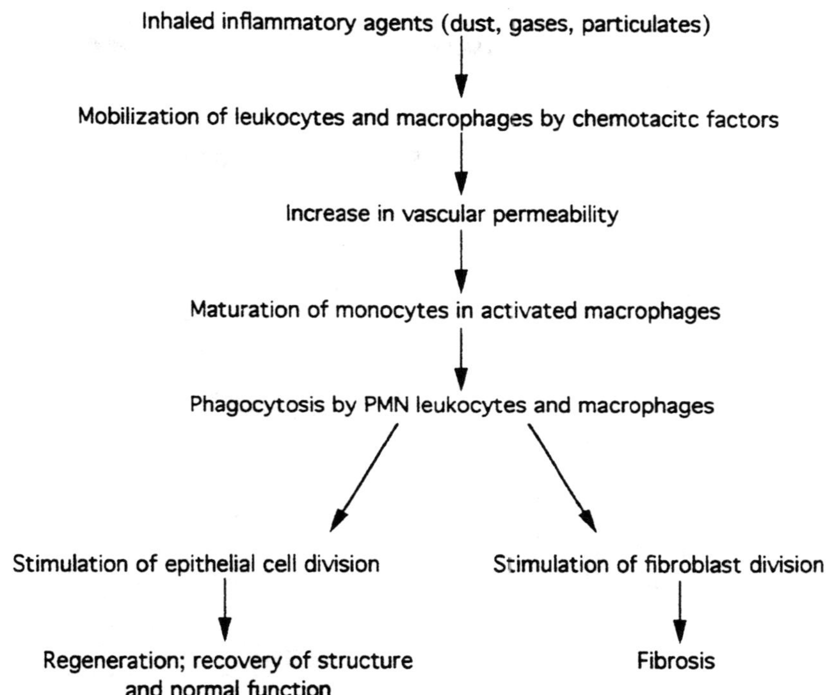

Figure 8.8. Events leading from exposure to an inflammatory agent.

protect against oxygen free radicals. Several experiments [31, 146, 191, 256] have demonstrated that following the exposure of rats to moderate doses of O_3 (2 - 3 ppm for 2 hours) or hyperbaric O_2 (85%), the surviving animals had increased levels of superoxide dismutase and glutathione peroxidase in their lungs. These surviving animals were also resistant to subsequent doses of O_3 (17 ppm for 3 hours) that would normally be lethal.

Similar increases in antioxidant enzymes, but not nonenzymatic antioxidants (vitamin C) are found in the blood of moderate smokers or miners. The chemiluminescence determinations on PMN leukocytes showed an increased activity, suggesting a heightened first line of defense. But this increased production of oxygen free radicals also provides the amplification factor that is responsible for the sequence of events leading to fibrosis and other pathological conditions. There is currently much interest in the possible defensive effect towards moderate exposure to chemical pollutants and dust from increased ingestion of vitamins C, E, and carotenes.

8.3.4 THE DOUBLE EDGED ROLE OF NITROGEN OXIDES

Generally people are somewhat familiar with SO_2 as a toxic gas in environmental pollution. Fewer are aware of the role of nitrogen oxides in the

Table 8.5
Biological actions of nitric oxide. After Mayorga [179] and Werner [292].

Organ/tissue	Benefit	Harm
Brain	Neurotransmitter, long-term effect	Neurotoxic
Stomach	Cytoprotective, reflex dilation, motility	Cytotoxic, possibly mutagenic
Pancreas	Insulin release	β-cell destruction
Blood vessels	Vasodilator, antithrombic	Reperfusion injury, anaphylactic shock
Leukocytes	Immune defense	Endotoxic shock

organism's metabolic activity.

Nitrous oxide (N_2O), popularly known as "laughing gas" because of its ability to cause euphoria, has been known for a long time. It was commonly used as an anesthetic from the middle of the nineteenth century, and continues in use today. In the past two decades, epidemiological studies have shown that serious health consequences may be associated with low levels of exposure to N_2O. The National Institute of Occupational Safety and Health, in 1987, recommended a maximal average level of exposure to N_2O of 25 ppm per procedure. The mean levels for the use of N_2O for procedures in private dental offices was more than 100 ppm.

Rowland [240] recently noted that women exposed to high levels of N_2O had significantly reduced fertility and described abnormalities of sperm and reduced fertility in male rats. The mechanism for this toxic effect is unknown, but is probably similar to the effect of smog.

Nitric oxide (NO) also has beneficial and harmful roles. Endogenous NO is synthesized from L-arginine by nitric oxide synthase. NO is now recognized as a chemical messenger. NO activates soluble guanylyl cyclase and forms the second messenger, nitric oxide-cyclic guanosine-3'-5'-monophosphate (CGMP). This pathway functions in vascular endothelial cells, which release NO, thereby increasing vascular smooth muscle cell CGMP levels and causing vasodilation. It can also be induced in vascular smooth muscle cells during sepsis, where NO formation may be responsible for the severe hypotension of septic shock [199, 289]. The NO signal transduction pathway is also present in epithelial cells, endocrine organs, and the nervous system as a neurotransmitter (table 8.5).

In activated macrophages, higher concentrations of NO can exert CGMP-independent effects that convert this signaling gas into a cytotoxic and mutagenic agent [19, 296]. Several recent experiments reported the serious implications of NO in specialized cells and tissues.

These discoveries concerning the dual role of NO are quite recent and much more research will be needed to understand the various aspects of this astonishing molecule.

CHAPTER 9

HEPATOTOXICITY

9.1 MECHANISMS OF ACTION OF CHEMICAL POLLUTANTS

Unlike the lungs, where the effect of chemical pollutants on the respiratory tract may alert the person to their presence, and consequently avoidance may be possible, penetration of chemical pollutants into the liver has no sensation. The exposure of the organism to chemical pollutants from food or water invariably involves the liver. This is the main organ involved in the metabolism and detoxification of xenobiotics (biotransformation).

Of course, not all chemical pollutants that penetrate the body reach the liver or are detoxified by that organ. Metals, for example, may be delayed in reaching the liver or may have specific effects on the brain or kidneys. But, for the majority of xenobiotic substances, the liver is the primary site of metabolism and detoxification. Due to its huge functional capacity, the liver is capable of facing long-term chronic exposures to chemical pollutants. There are, of course, limits to this capacity, which vary between individuals.

The enormous resilience of the liver endows the organism with a mechanism for adaptation that can raise the tolerance threshold through enzymatic induction. A common example is ethyl alcohol, where adaptation is frequently demonstrated. Therefore, the clear determination that a chemical pollutant is involved in liver failure is often difficult. Most chemicals determined to be hepatotoxic have had their effects discovered because of occupational exposure (metals, organic chemicals) or through animal experimentation [4, 49, 259, 264]. Unfortunately, extrapolation from animals to humans is not always possible.

In the lungs, the effect of a chemical pollutant is increased by tobacco smoke. In the liver the same effect occurs with alcohol. The massive functional capacity of the liver and its adaptability to noxious substances is a great advantage. However, in some ways, this advantage may be off set by the inability to sense the threshold level has been exceeded until clinical symptoms appear. As was seen in figure 6.1, this is a late point in the exposure and much damage at the molecular, cellular, and functional levels has been done before this point is reached. Most xenobiotics that penetrate the body reach the liver and are metabolized. But some of them are transformed into dangerous active metabolites, or contribute to the activation of oxygen.

This explains the great amount of enzymatic and nonenzymatic antioxidant systems located in the liver: they are necessary to provide resistance to oxidative reactions. In addition to exposure to chemical pollutants, the liver can be damaged by toxic compounds, bacterial toxins, metals, drugs, etc., either alone or in combinations. Ethanol, lead, and nitrosoamines are the toxic compounds that most commonly challenge the liver and its functions.

All xenobiotics damage the liver through one of two basic mechanisms:

A. *Cholestasis*: the clinical situation in which the passage of bile is reduced. This leads to the accumulation of bilirubin in hepatocytes and Kupfer cells. Blockages and lesions develop in the biliary channels and a mechanical jaundice develops. The mechanism of cholestasis proceeds with damage to the biliary secretory system in the liver, followed by the dilation of the biliary channels, structural modification of the microvilli, and endoplasmic reticulum by modification of the protein/phospholipid ratio.

Physiologically, the flow of bile is regulated by two mechanisms that depend on the transport of ions, water, and biliary salts driven by ATPase activity. Ion pumps are located in the microvili of the biliary channels and the transport of ions precedes that of bile salts. The osmotic force produced by the transport of ions and bile salts provides the necessary energy.

Chemicals, especially steroid hormones, affect the bile flow by modifying the structure of micelles formed by bile salts and phospholipids. These polyionic micelles or particles are formed in the endoplasmic reticulum of hepatocytes and contain cholesterol, bile acids, and phospholipids. Exogenous chemicals act at this point by interfering with the conversion of cholesterol to bile acids or by modifying the structure of the micelles leading to the formation of clots that obstruct the biliary channels. These biochemical modifications lead to a change in the concentration of ATPase, g-glutamyl transpeptidase, sialic acid, phospholipids, cholesterol, and bilirubin.

Other than steroid hormones, some drugs, like phenothiazine (chlorpromazine), act according to this mechanism. In food, there are many compounds present that induce the gall bladder to contract stimulating the flow of bile. These compounds include fats and oils, peptones (protein derivatives), phenols, naphthols, aldehydes, chlorinated derivatives, ethers, etc. The interaction of these choleretic compounds with chemical pollutants, infections, and alcohol intoxication promotes cholestasis to various degrees. Early clinical manifestations of cholestasis are common among people sensitive to chemicals. A diet free of additives and without fried foods should improve the condition [253, 264].

B. *Hepatocellular lesions*: damage to the liver. Experimental intoxication with carbon tetrachloride (CCl_4) has become a classical model for the production of parenchymal hepatotoxicity.

Damage to hepatocytes varies according to the nature of the attacking compound. The morphologic aspects are similar to those observed with chronic hepatitis: hepatic cellular necrosis, hyperplasia of endoplasmic reticulum, and cholestasis. Intoxication with carbon tetrachloride or ethanol also involves fatty infiltration and the inflammation of portal channels.

It should be mentioned that the involvement of chemical pollutants in hepatotoxicity is very complex, involving simultaneous damage to other organs (lead, phenothiazines, nitrosoamines) or proceeding by both of the above mechanisms. The equilibrium between toxic agents and detoxification systems is greatly influenced by diet and other possible hepato-protective factors.

9.2 CARBON TETRACHLORIDE; A MODEL FOR HEPATOTOXICITY

Carbon tetrachloride hepatotoxicity was originally seen mostly as a result of occupational exposure. However, it is now used in many experimental studies as a reference model. Experimental intoxication with carbon tetrachloride is not simple, requiring considerable effort to do correctly, and frequently presents surprises. The rat, the animal commonly used for these experiments, is remarkably resistant to this agent. In spite of a great deal of literature on the topic, it is only in recent years that a clear understanding of the sequence of steps leading to hepatotoxicity has been accepted [159, 163, 230].

The sequence of biochemical, biological, and clinical events involved in intoxication with carbon tetrachloride clearly illustrates the complexities and difficulties already mentioned. The first step, solvent metabolism, is a double-edged process. The organism attempts to remove the noxious compound through biotransformation, but instead forms free radicals, specifically the trichloromethane ($CCl_3^{\bullet}$) and trichloromethylperoxyl ($CCl_3OO^{\bullet}$) radicals. The formation of these free radicals takes place in the smooth endoplasmic reticulum by cytochrome P_{450}.

Several studies [7, 105, 137] have shown that 24 hours after an acute carbon tetrachloride intoxication, glutathione, cytochrome P_{450} and some other metabolic enzymes are significantly decreased. This is in agreement with the events presented in table 9.1. Most of these start as early as 5 minutes following and acute exposure, but peroxidized phospholipid decomposition can only be detected after 1 - 12 hours [23, 213, 246, 264]. The biochemical events that occur during carbon tetrachloride intoxication have been studied for over 20 years. It was an Italian team from the University of Siena that showed that carbon tetrachloride radical formation and covalent binding to liver proteins proceeds even under anaerobic conditions. Under aerobic conditions, peroxidation of membrane lipids occurs in addition to the free radical formation, due to oxygen activation.

Another problem, closely related to peroxidation, was the distant action or propagation of oxygen activation. As described in figure 9.1, peroxidation is a process that proceeds like a chain reaction by generating its own catalyst. The carbon tetrachloride free radicals are very reactive and exhibit too short a life span to be involved in long distance actions. Neither are lipid peroxides implicated, based on membrane structural and permeability factors. The Italian team eventually demonstrated that the oxidative damage caused by carbon tetrachloride intoxication is perpetuated by aldehydes and ketones that result from peroxide decomposition. These aldehydes possess mitogenic and cytotoxic properties, being able to damage structures and alter biological

Table 9.1
The main steps in carbon tetrachloride intoxication

Phase I: <u>Solvent metabolism in the liver</u>
1. Involvement of MEOS in endoplasmic reticulum
2. Production of CCl_4 free radicals ($CCl_3^\bullet$)
3. Covalent binding of $CCl_3^\bullet$ to proteins and lipids
4. Reaction of $CCl_3^\bullet$ with polyunsaturated fatty acids (PUFA) of the endoplasmic reticulum
Phase II: <u>Peroxidatin of PUFA</u>
5. Peroxidation of PUFA (possible iron dependent)
6. Decrease in content of reduced glutathione
7. Decomposition of lipid peroxides, increased aldehyde formation (MDA), and their elimination by expired air (pentane, ethane)
8. Inactivation of some enzymes (glucose-6-phosphatase)
9. Propagation of diffusion of lipid peroxides and aldehydes
10. Inactivation of some membrane bound enzymes, cellular membranes become hyperpermiable, hydrolases released with cell lysis
11. Covalent binding of aldehydes to proteins, inhibition of protein synthesis
Phase III: <u>Morphologic modifications</u> and <u>clinical manifestations</u>
12. Fatty infiltration of liver resulting in increased content of triglycerides
13. Apperance of necrosis and cellular hyperplasia
14. Cholestasis and related alterations in physiology

functions [84, 113, 201].

Dialyzable compounds resulting from the peroxidation of hepatic microsomes had already been shown to have hemolytic properties, inactivate proteins, and oxidize glutathione and free SH groups. These diffusible compounds were identified as aldehydes: 4-hydroxy-trans-2, 3-nonenal (4HNE, CH_3-$(CH_2)_4$-CH(OH)-CH=CH-CHO, 95%), 4,5-dihydroxy-trans-decenal (4,5-DD, 5%), and traces of 4-hydroxy-2, 3-decenal. These studies, involving high performance liquid chromatography and mass spectrometry clearly established their structures [48, 194, 230, 300]. 4HNE was synthesized and a close correlation was found between the action of this compound and the subsequent effects of peroxidation (table 9.2).

                    R. Olinescu, T. Smith and J. Hertoghe

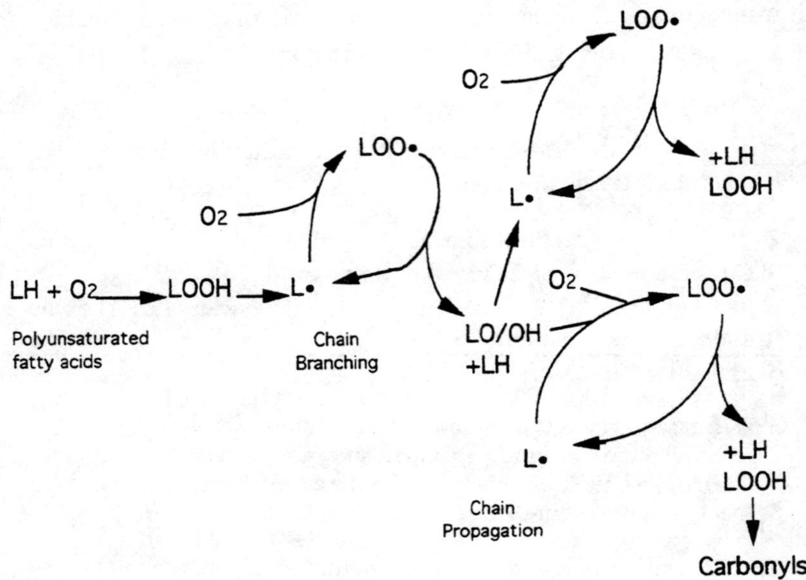

Figure 9.1. Chain reactions leading to the formation of fatty acid hydroperoxides. LH is a fatty acid (PUFA), while L• is a related free radical and LOOH is a hydroperoxide.

Table 9.2
Biochemical effects of 4-hydroxy-2,3-nonenal (4-HNE)

Process	% Inactivated
Microsomal glucose-6-phosphatase	86
Microsomal cytochrome P450	47
Microsomal NADPH-cytochrome reductase	17
Reticulocyte protein synthesis	53
Loss of hepatocyte viability	51
Microsomal hemolytic activity	—
Calcium pump (ID50 = 12 μm)	—
Chemotactic activity (0.001 - 0.1 μm)	—

These α or β unsaturated aldehydes are very reactive with nucleic acids, glutathione, and other thiols due to their electrophilic nature. Given that the concentration of glutathione in liver is around 6 mM, a half-life of 3 minutes was calculated for 4HNE. For cellular metabolic reactions, this represents a relatively long life span, sufficient to propagate the effects of peroxidation throughout the cell and to its nearby environment. Great permeability of 4HNE is assured by its having a hydrophilic segment (-CHOH-CH=CH-CHO) and a lipophilic segment (CH_3-$(CH_2)_4$-). Given this structure, aldehydes like 4HNE are capable of crossing all biological barriers.

As mentioned in chapter 4, malondialdehyde (MDA) is a natural product of lipid peroxide decomposition [92]. Among all the aldehydes resulting from the decomposition of lipid peroxides, MDA is sufficiently hydrophilic to leave the cell and accumulate in the blood. It is calculated that following intoxication with carbon tetrachloride or bromo-trichloride ($BrCCl_3$), a concentration of 1 - 4 mM MDA will be found in the blood [49]. This is sufficient to link all cytopathologic effects throughout the body to the exposure to the solvent.

As described in table 9.3, chemical-induced hepatotoxicity involves 3 types of compounds. These compounds cause damage according to different mechanisms, but all share the common factor of oxygen activation.

A second thing to consider is the drop of glutathione that precedes lipid peroxidation. Several scientists confirm that a decrease of glutathione must occur prior to the peroxidation of membrane lipids [40, 117, 155, 244]. It has been demonstrated that for each organ there is a critical threshold of glutathione (about 2 mM), below which peroxidation can take place [40, 181, 220]. This explains the protective or therapeutic effect of compounds that provide free SH groups, such as N-acetylcysteine. Glutathione is not only involved in the second phase of biotransformation (mercapturic acid formation), but also in protecting the functional and structural integrity of liver cell, and as an essential component of antioxidative systems (substrate for glutathione peroxidase).

A clear-cut experiment was done by Benedetti [23] that showed the results of intoxication with bromobenzene and the significant protection afforded by Trolox (a mixture of vitamins C and E) (see figure 9.2). Experimental animals were separated into 3 groups according to the degree of

Table 9.3

Mechanism of chemically-induced hepatotoxiciy for the three main groups of compounds involved.

CCl_4 and $BrCCl_3$	Bromobenzene and acetaminophen	Paraquat, diquat, and methadone
Drug metabol. → alkyl radicals Covalent Lipid binding peroxidation	Drug metabol. → electrophilic intermediates ↓ GSH depletion	Drug metabol. → nonalkylating metabolites ↓ active oxygen species

Other alkylating agents:
  $CHBr_3$ (bromoform)
  CHI_3 (iodoform)
  CH_2I_2 (methyl diiodide)
  $CH_2=CHCl$ (2-monochloroethylene, vinyl chloride)
  CF_3-$CHBrCl$ (halothane, fluothane)
  $CHFCl$-$CFOCF_2H$ (ethrane)
	Other GSH depleting agents:
	  Diethylmaleate
	  CHBr-CHBr (1,2-dibromoethane, ethylene dibromide)
	  $CCl_2=CH_2$ (1,2-dichloroethylene, vinylidene chloride)
	  $CCl_2=CClH$ (1,1,2-trichloroethylene, trielin)
	  CHCl=CHCl (1,2-dichloroethane)
	  $CHCl_3$ (chloroform)
	  CH_3I (methyl iodide)
	  1,2-dichloro-4-nitrobenzene
	  Allyl alcohol
	  Acrylamide
	  Acrylonitrile
	  Benzyl chloride
	  Chloroethanol
	  Cinnamaldehyde
	  Cyclohex-2-en-1-one
	  Dichlorvos
	  a-Hexachlorocyclohexane
	  1-Nitrobutane
	  Parasorbic acid
		Other redox cycling compounds:
		  Adriamicin, Doxorubicin
		  Mitomycin C, antimycin
		  β-Lapachone
		  Alloxan
		  Bleomycin
		  Nitrofurantoin and other aromatic nitrocompounds
		  Hydrazine and its derivatives (phenylhydrazine)
		  6-Hydroxydopamine

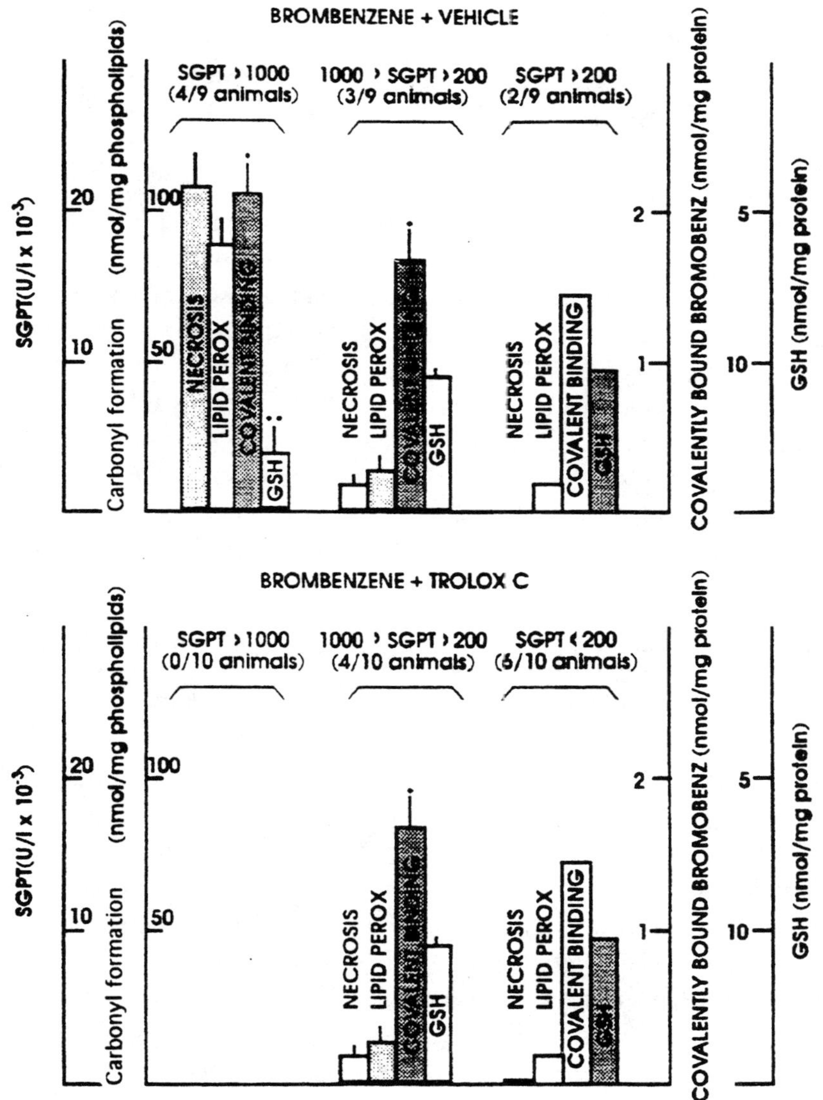

Figure 9.2. Liver necrosis (determined from serum glutamate-pyrivate transaminase, SGPT), liver peroxidation (amonunt of carbonyl formation in liver phospholipids), covalent binding of [14C] bromobenzene metabolites to liver protein and hepatic GSH content in mice intoxicated with bromobenzene (13 mmoles/kg body weight, by mounth) and given either Trolox C or the vehicle (propylene glycol) alone. Trolox C was given intraperitoneally at a dose of 270 μmoles/kg body weight at 9 and 13 hours after bromobenzene dosing. The animals were sacrificed 18 hours after bromobenzene dosing. The data are presented as means ± SEM. From Benedetti [23].

intoxication manifested (determined by measuring glutathione peroxidase and transaminase) and by liver necrosis. Covalent binding of bromobenzene was the only parameter not influenced by antioxidants. This was not a surprise since this process does not depend on oxygen activation. A diet rich in proteins (a source of glutathione) also offers a significant amount of protection compared to a diet based on glucose and vegetables. This experiment clearly demonstrates the different paths of the chemical for binding and necrosis, with the later depending on peroxidation.

It has been demonstrated that in any metabolic process where a decrease in glutathione results, peroxidation proceeds, provided the lower threshold for glutathione is crossed. This also implies the equilibrium between glutathione regeneration and consumption favors consumption. An excessive rate of glutathione oxidation will allow the accumulation of GSSG in the cell, followed by its diffusion out of the cell where it can be measured as an indicator of oxidative stress [39, 48, 181].

The increase of the GSSG concentration in the extracellular fluids will induce:

- Increased cell membrane permeability
- Inactivation of enzymes involved in transmembrane transport: ATPase, translocase, phosphokinases
- Alteration of the calcium pump

That hepatotoxicity depends on the modification of glutathione and on peroxidation was proven experimentally by B. Chance and E. Cadenas of Rockefeller Institute, and H. Sies and L Flohe of the University of Tubigen. They used special equipment to dynamically measure the development of hepatotoxicity induced by a chemical and the *in situ* formation of GSSG as well as chemiluminescense [53, 214].

The protection or partial reversal of chemically-induced hepatotoxicity (carbon tetrachloride, bromobenzene, paracetamol) has been obtained with:

- N-acetylcysteine, cystamine (favors glutathione regeneration)
- Vitamins C and E (antioxidants protecting from peroxidation)
- Selenite (1 ppm, required for the activity of glutathione peroxidase)

- Tocopherols (vitamin E), carotenes, purines, or phenothiazines (free radical scavengers)

The administration of stimulators of biotransformation activity, such as alcohol or phenobarbital, produces unpredictable effects that range from partial inhibition to increased potency of the challenging agent due to interactions (chapter 3).

Table 9.4
World production and consumption of ethyl alcohol (1988)

Production (million gallons / year)		Consumption (gallons per person / year)	
Italy	2,000	France	23
France	1,700	Portugal	20
Spain	900	Italy	19.5
USSR	820	Argentina	19
Argentina	620	Spain	16
USA	435	USSR	10
Portugal	250	USA	3
Yugoslovia	200	Finland	2

9.3 ETHYL ALCOHOL

Ethyl alcohol, consumed in various forms, is the most frequent xenobiotic encountered by the body, exceeding the use of tobacco in both production and use. While campaigns to decrease tobacco smoking have had some success, campaign against drinking alcohol have had little success. Alcohol consumption has both biochemical and clinical consequences as well as social and economic effects. Relatively recent data on the production and consumption of alcohol in the world are presented in table 9.4.

The widespread use of alcohol around the world has generated a public reaction and challenge for scientist. At least one international congress a year is dedicated to aspects of drinking alcohol. The International Commission for the Environment (ICPEMC) in 1987 included ethanol among mutagens, mainly due to its metabolic by-product, acetaldehyde [113, 251].

Scientists have been impressed while studying the biochemical consequences of alcohol by the variety of adaptations seen in humans as well as the diversity of reactions involved. Due to polymorphism of some enzymes, especially aldehyde dehydrogenase, populations from Asia and the Southern Hemisphere experience greater difficulties with the handling of acetaldehyde, the main product of ethanol metabolism.

The differences in the metabolism of alcohol begin at the most basic level. Ethyl alcohol has a caloric value of 7 cal/g. This places it between the caloric content of carbohydrates (4 cal/g) and fat (9 cal/g). Consequently, 0.5 liter of 43 proof liquor provides half the needed calories for a day. But the calories provided by alcohol are "empty calories" because they are not associated with any nutrients like protein, minerals, or vitamins. At the same time, the appetite is partially satiated because of the calories and malnutrition becomes a risk, especially in alcoholics.

Ethyl alcohol is easily absorbed from the gastrointestinal tract, but with great individual variation in rate. Its absorption is also influenced by numerous factors such as the type of liquor, the amount consumed and over what time span, and the presence or absence of food.

In 1919, Mellamby found that ethyl alcohol is metabolized in the liver at a constant rate of 5 - 7 g/hour, but more recent work has discovered quantitative modifiers of this rate. The amount of alcohol that appears in the blood at a given time after ingestion varies for the drink (for a constant sized drink): vodka > whiskey (43 proof) > liquor (22 proof) > wine (12 proof) > beer (6 proof). If food is eaten with the drink, the amount in the blood varies with: empty stomach > proteins > lipids > carbohydrates.

Unlike true nutrients, ethyl alcohol is not stored in tissues, so the body can only be rid of it through oxidative metabolism. Unlike fats and most carbohydrates, which can be oxidized in nearly all tissues, alcohol can only be oxidized in the liver. Only a small amount of alcohol (less than 10%) is eliminated through exhalation. The organ specificity of alcohol oxidation helps explain the extent of its harmful effect on the liver.

Another property of alcohol involves its self-metabolic acceleration as an adaptive response of the body. Prolonged low consumption of alcohol leads to an increased rate of ethanol oxidation in both experimental animals and man. This is called "metabolic tolerance" of alcohol [22, 106, 161, 282].

As seen in figure 9.3, ethyl alcohol can be metabolized enzymatically through three reactions. The major reaction is catalyzed by alcohol dehydrogenase. At low concentrations of alcohol, the reaction mediated by

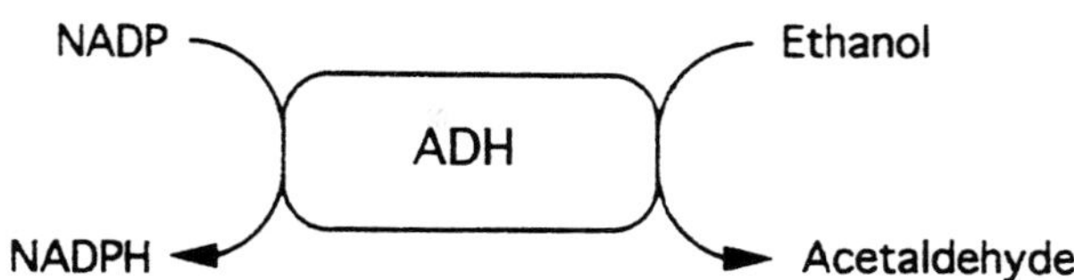

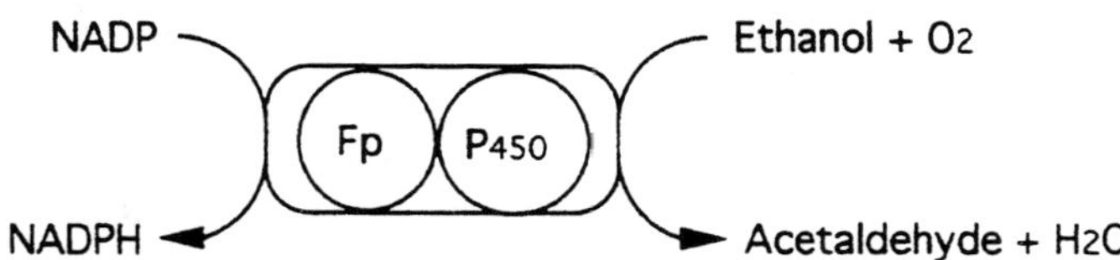

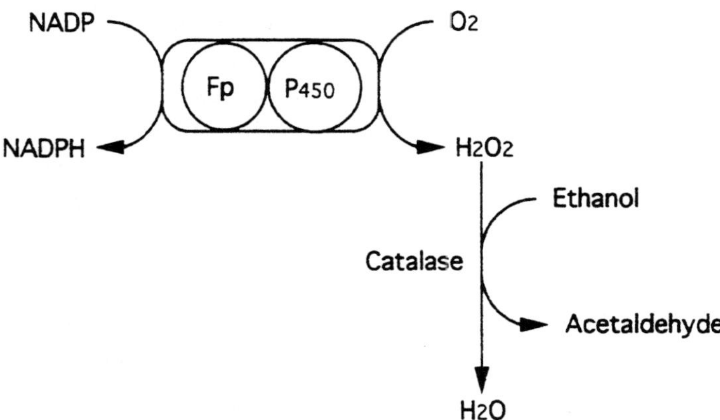

Figure 9.3. Mechanisms and metabolic interrelationships of microsomal ethanol oxidation. A: Primary pathway, oxidation of ethanol by alcohol dehydrogenase (ADH). B and C are secondary pathways. B: Direct oxidation of ethanol by cytochrome P_{450}. Fp = flavoprotein. C: NADPH-dependent hydrogen peroxide production with the subsequent metabolism of ethanol by peroxidase activity of catalase.

zinc-dependent alcohol dehydrogenase accounts for nearly the entire metabolism. Ethanol oxidation by all three enzymatic reactions produces the same metabolite: acetaldehyde. Acetaldehyde is subsequently broken down by aldehyde dehydrogenase, also found in the liver.

Ethanol oxidation is stimulated by artificial electron acceptors, glucogenic precursors and fructose. These compounds stimulate ADP formation from ATP, increasing the electron flux in the mitochondrial respiratory chain, thus enhancing the reoxidation rate of NADH, which is required for the metabolism of alcohol.

$$\text{Ethanol} + \text{NAD} \xrightarrow{\text{alcohol dehydrog.}} \text{Acetaldehyde} + \text{NADH} \qquad (1)$$

$$\text{Acetaldehyde} + \text{NAD} \xrightarrow{\text{acet. dehydrog.}} \text{Acetate} + \text{NADH} \qquad (2)$$

Therefore, the rate limiting step for alcohol oxidation is ultimately the rate of electron flux in the mitochondrial respiratory chain, a process controlled by the supply of NAD, because ethanol intoxication produces a decrease in the NAD/NADH ratio between the mitochondria and the cytoplasm.

At high ethanol concentrations alcohol metabolism is about evenly divided between dehydrogenase and catalase. Experimental evidence concerning cytochrome P_{450} involvement in ethanol metabolism is still under discussion. The peroxidative reaction of catalase probably represents a supplementary pathway for the detoxification of a large number of small molecules, including ethanol, methanol, nitrite, and formate.

Alcohol dehydrogenase has considerable polymorphism: 9 genotypes and 15 isoenzymes. The presence of a large number of atypical phenotypes of this enzyme (3 - 20% for Caucasians; 80 - 90% for Asians) explains the great variation in alcohol tolerance for races and individuals [168, 282].

Apart from the metabolic complications of excess calories and of adaptive changes in microsomal activity, heavy alcohol consumption has a direct toxic effect on liver tissue. A central role in the toxicity of alcohol may be played by acetaldehyde, a product of three enzymatic pathways (figure 9.2). Acetaldehyde is extremely reactive and affects most tissues of the body. Most acetaldehyde is converted into acetate by liver mitochondria (reaction 2), but some escapes into the bloodstream. All the enzymes (figure 9.2) become saturated when the liver content of alcohol reaches a certain point and remain saturated until the alcohol level drops. The acetaldehyde plateau is significantly higher in alcoholics than it is in nonalcoholics, even when the same amount of alcohol is given to both groups and the same level of blood alcohol has been reached. This abnormally high acetaldehyde level in alcoholics could result from a more rapid conversion of alcohol to acetaldehyde. The alcoholic may, therefore, be the victim of a vicious cycle: a high acetaldehyde level impairs mitochondrial function in the liver, acetaldehyde oxidation is decreased, more acetaldehyde accumulates and causes further liver damage.

Acetaldehyde does not only affect the liver, but other tissues as well: heart, muscle, blood vessels in the extremities. Damage to these tissues, along with tolerance, characterizes alcohol addiction. Most of this damage can be attributed to acetaldehyde. Dependence manifests itself by a state of extreme discomfort, often accompanied by physiological disturbances such as tremors and seizures upon withdrawal of the drug. It appears that acetaldehyde affects the metabolism of neurotransmitters in the brain.

If heavy drinking continues, reversible fatty liver develops in most people. This can evolve into more severe and irreversible liver damage and eventually cirrhosis. In table 9.5, some histological and biochemical modifications are presented that follow chronic intoxication of rats with ethanol. It can be observed that the modifications in the liver only became significant after 2.5 months. This experiment demonstrates the great functional resistance of liver and the parallel but opposite changes in triglycerides and glutathione. Blood lipid peroxides reached a significant level only after there was a significant drop in liver glutathione.

Lieber [161] observed a progressive increase in fatty liver in man using a thin needle biopsy. The volunteers were fed a low fat diet and had 6 drinks per day for a total of about 10 ounces of 86 proof alcohol. The condition of fatty liver also brings about changes in the ultrastructure of the hepatocytes, especially the mitochondria and endoplasmic reticulum. This occurs with a rather moderate ingestion of ethanol (80 -90 mg /100 ml), which did not result in any clinical signs of intoxication. It is not necessary to become drunk for liver damage to occur.

The first stage of ethanol-induced injury in humans is alcoholic hepatitis, in which decreased liver cell function leads the death of cells, inflammation, and a mortality of 10 - 30%. The final stage is cirrhosis, when fibrous scars disrupt liver architecture and give rise to a number of potentially fatal complications. However, it takes from 5 to 25 years of steady drinking for a human to develop cirrhosis.

Another aspect of alcohol metabolism deals with the increased capacity of an alcoholic's liver to metabolize various drugs. Once adaptation has occurred, alcoholics require larger doses of many drugs, including sedatives. That is why simultaneous drinking and taking tranquilizers is particularly dangerous. The presence of alcohol interferes with the liver's capacity to

Table 9.5

Histological and biochemical modifications in rat liver following chronic intoxication with ethanol. After Olinescu et al. [213].

Time (months)	Histologic stage	Lactate/Pyruvate ratio	Triglyceride (mg/g)	Glutathione (mg/g)
Initial	0	12±3.4	3.2±0.8	2.8±0.5
1	1	14±3.5	4.1±0.6	3.2±0.3
1.5	1	21±5.8	5.6±1.2	2.6±0.2
2	2	34±9.4	7.5±1.3	1.9±0.4
2.5	3	38±6.8*	10.3±2.4**	1.6±0.2*
3	3	46±11**	14.6±3.7**	1.3±0.2*
3.5	3	54±9	18.7±4.2**	0.8±0.3**
4	4	65±10	22.6±5.2	0.5±0.1**
Control	1	19±7	9.7±1.8	1.9±0.8

Triglycerides and glutathione were expressed as mg per gram of fresh liver. The histological modifications were classified into groups according to the following scheme: 1 = minimal changes, sinusoid dilation; 2 = mild degeneration, disruption of lobular pattern; 3 = more extensive degeneration, swelling of hepatocytes, lymphocyte infiltration, fatty degeneration; 4 = marked degeneration, periportal necrosis and hydropic degeneration. Statistical values of $p<0.05$ are indicated by * and values of $p<0.01$ are indicated by **. These modifications were obtained after a diet rich in lipids and poor in proteins and lipotropic factors.

inactivate the drug.

It took almost 20 years of intense research to demonstrate the formation of free radicals during acute or chronic intoxication with alcohol. Ethanol intake (acute or chronic) elicit complex hepatic disturbances that include the formation of hydroxyethyl radicals ($CH_3C^\bullet HOH$ and $C^\bullet H_2CH_2OH$) and the hydroxyl radical (OH$^\bullet$). An oxidative stress condition is also favored by increased availability of free iron ions, formation of ethanol-inducible cytochrome P_{450} (a form called CYP2E1), the involvement of xanthine dehydrogenase in the metabolism of acetaldehyde and leukocyte activation from alcoholic hepatitis. These complex modifications contribute to and explain the irreversible progression towards cirrhosis with chronic alcohol intoxication.

Atherosclerosis and coronary heart disease (CHD) have been linked to the consumption of dietary fat, especially saturated fats. But, a balanced diet rich in foods of plant origin can significantly retard the development of CHD. This relates to the "French paradox." In most countries, a high intake of saturated fats is strongly correlated with high mortality from CHD, but not in some regions of France. Alcohol intake in moderation (1 glass of wine a day) may protect from CHD, but the risk increases at higher levels of consumption. Red wine contains several phenolic substances (flavonoids, catechins, soluble tannins). The antioxidative properties of these compounds may delay the onset of athersclerosis by reducing and down-regulating thrombotic tendencies [52].

9.4 NITRITES AND NITROSOAMINES

Nitrates and nitrosoamines have been on lists of chemical pollutants for approximately 20 years. This period is too short for a clear-cut demonstration of the biological and clinical implications following man's exposure. These compounds are incriminated in chemically-induced carcinogenesis.

Nitrosoamines possess the general formula of R_1R_2-N-N=O and have a moderate hepatotoxic activity. Depending on the substitutions (R_1R_2), nitrosoamines have a lethal dose ranging from 20 - 5,000 mg/kg. Following acute intoxication in rats, hepatic and pulmonary lesions have been detected. Following chronic intoxication (micrograms daily) in all tested animals, cancer of the liver and lungs occurs [144, 185]. Among 100 nitrosoamines tested, 65 exhibited carcinogenic potential under experimental conditions. In humans the effect of nitrosoamines is still unclear.

Table 9.6 lists the correlation between nitrites and nitrosoamines and tumors in various countries. These conclusions correlate with experimental findings in animals chronically exposed (micrograms daily) to these compounds. Pesticides like Ziram, Atazine, and Ferbam or drugs like chlordiazoepoxide, aminopyrine, and methadone produce nitrosoamine related derivatives during their metabolism [38, 49, 52, 226].

The biochemical transformation of these compounds is not completely understood. Transformation of nitrates into nitrites only takes place at a pH greater than 5, while the biosynthesis of nitrosoamines from nitrites only takes place at a pH lower than 3. Therefore, conditions are only suitable in the stomach to convert nitrites to nitrosoamines:

Table 9.6
Implication of nitrosoamines in human carcinogenesis

Population	Source	Compound	Location
Bantu (Africa)	Solanum incam.	Nitrosoamines	esophagus, liver
Zambia (Africa)	Beverage (plants)	Nitrosoamines	esophagus, liver
Normandy	Cavados (drink)	Nitrosoamines	esophagus, liver
England	Tap water	Nitrates	stomach
Columbia	Tap water	Nitrates	stomach
China	unknown	Nitrosoamines	Nasophagingeal
USA	unknown	Nitrosoamines	Pancreas
USA	Smoked or fried meat	Nitrates	Stomach, liver
Japan, China	Fertilizer	Nitrosoamines	Stomach, liver

$$\text{Nitrites} + H^+ \rightarrow \text{Nitrous acid (HNO}_2) + H^+ \rightarrow N_2O_3 \qquad (3)$$

$$R_2NH + N_2O_3 \rightarrow R_2N\text{-}NO + HNO_2 \qquad (4)$$

Therefore, the amount of oxidized nitrogen compounds produced in the stomach or during food storage will depend partly on the nitrosylation kinetics, i.e., the rate equation and their constants:

$$\text{rate} = K_2 \, [\text{amine}] \, [\text{nitrite}]^2 \qquad (5)$$

As determined from kinetics studies, the rate of nitrosylation increases as the basicity of amines decreases. Fish contain relatively large amounts of dimethylamine. The nitrosylation of secondary amines (R_2NH) occurs during fermentation or cooking (reaction 4). The rate of reaction (4) is maximal at a pH value of 3.4. Such a pH is only found in the stomach of people with aclorhydria, where a microbial flora is able to synthesize nitrosoamines. In such people, a higher frequency of gastric cancer is found [240]. Bacteria like *E. coli* and *Staphylococcus aureus* may also act in the gall bladder. *In vivo* nitrosylation may also occur in the large intestine, the infected bladder, and the infected vagina. In the case of the large intestine, *Streptococcal* bacteria behave as nitrosylation catalysts. Urine becomes acidic when infected with *E. coli*. This could reduce nitrate to nitrite, which could then react enzymatically with secondary amines normally present in urine (dimethylamine, piperidine, pyrrolidine; which mainly results from the metabolism of intestinal bacteria).

Nitrosylation is favored by the presence of thiocyanates (NCS^-). The greatest amount of thiocyanate is found in saliva (6 mM) and gastric juice (0.7 mM), especially in smokers. The normal level of nitrites in saliva is 10 ng/liter [127, 171, 245]. Thiocyanates result from nitrites and may be coupled to reaction (3) and (4). Cigarettes contain 0.2 - 1% nitrites, which, following combustion, will produce 84 ng of dimethylnitrosoamine per cigarette.

Thiocyanates increase the nitrosylation of aniline, morpholine, and sarcozine by the following reactions, but only when the pH is 2:

$$NCS^- + HNO_2 \rightarrow ON\text{-}NCS + H_2O \tag{6}$$

$$ON\text{-}NCS + R_2NH_2 \rightarrow R_2N\text{-}NO + H^+ + NCS^- \tag{7}$$

Fortunately, these reactions are difficult to reproduce *in vivo*, even in experimental animals. The carcinogenic effect of nitrosoamines takes place by the alkylation of purines and pyrimidines, components of nucleic acids. Dimethylnitrosoamine (DMN) behaves as a methyl donor towards cytosine and guanine, resulting in the formation of 3-methylcytosine and 7-methylguanine. Once alkylated, the modified nitrogen bases induce errors into the genetic code and subsequently into the amino acid sequence of proteins.

A team from the Massachusetts Institute of Technology demonstrated another mechanism of alkylating bases starting with nitrosopyrrolidine (NP), a chemical present in smoked meat and tobacco smoke. NP has a structural resemblance to the natural amino acid proline. In liver, NP is enzymatically activated by o-hydroxylation.

$$R\text{-}CH_2\text{-}N(NO)\text{-}CH_2\text{-}R' \xrightarrow{\textit{hydrolysis}} R\text{-}CH_2\text{-}N(NO)\text{-}CHOH\text{-}R'$$

$$R\text{-}CH_2\text{-}N{=}N\text{-}OHRCH_2\text{-}N{\equiv}N \xrightarrow{\textit{DNA}} CH_3\text{-}DNA \qquad (8)$$

The final compound and some intermediates resulting from reaction (8) are mutagenic to *S. typhimurium*. The strongest alkylating intermediate is alcanediazotic acid ($R\text{-}CH_2\text{-}N{=}N\text{-}OH$). Fortunately, compounds are also present *in vivo* that competitively inhibit this process. Thus, spermidine (a polyamine), distamicin (an antibiotic), and elagic acid (a plant polyphenol) effectively inhibit the nitrosylation of DNA.

Table 9.7

Variation in biochemical parameters following intoxication of rats with DMN. After Slater [264].

Treatment	Transaminase GPT (units/dl)	Liver necrosis	Vitamin E (μg/dl plasma)
Oil + saline	150±36	0	130±7
Vitamin E + saline	72±29	0	290±10*
Selenium + saline	138±44	0	250±12*
Oil + DMN	521±204	4	150±8
Vitamin E +DMN	283±77*	3	310±7*
Selenium + DMN	1210±714*	4	180±7*

GTP = Glutamat Pyruvate Transaminase
* Significance difference compared to control, p<0.01.

The rate of nitrosylation is significantly increased by thiocyanates, halogens, and formaldehyde and inhibited by ascorbic and gallic acids, tannins, glutathione, and cysteine. By feeding ascorbate (23 g/kg) to rats, Meyskens [185] succeeded in inhibiting the process of DMN-induced hepatoma or pulmonary adenoma formation.

Ascorbic acid reacts with nitrites in the acidic medium present in the stomach. Ascorbate is added to nitrite-preserved caned meats to increase the formation of nitric oxide, which is responsible for the pink color of nitrosomyoglobin in meat. It has also been demonstrated that ascorbate inhibites DMN production in frankfurters treated with high concentrations of nitrites. Meyskens [185] showed that ascorbic and tannic acids, cysteine, and sodium sulfite, in concentrations in the range of 10 - 20 mM completely inhibits the nitrosylation of piperazine and morpholine at pH2.

Experimental evidence shows that during the metabolic activation of nitrosoamines in liver, free radicals are produced. The formation of free radicals during the metabolic activation of carcinogens has been demonstrated for only a few compounds (e.g., polyaromatic hydrocarbons). This is easily explained by the effectiveness of vitamins C and E in scavenging the free radicals formed during *in vivo* activation [4, 29, 126]. The protective effects of some antioxidant systems is shown in table 9.7. Liver necrosis was quantified using the same standards as described in table 9.5, with * indicating a statistical significance of p<0.01. Some antioxidants, such as 0.02% vitamin E, added to the diet protect against the hepatotoxicty of DMN (30 mg/kg) as confirmed by glutamate pyruvate transaminase (GPT) measurement and histological examination [51, 219]. The most interesting effect was attributed to selenium, which did not protect from the effect of intoxication with DMN. The negative result with respect to selenium may be attributed to a phenomenon of cross-induction, in which DMN is metabolized by other pathways in addition to MEOS [55, 70, 128].

Based on such studies, the use of nitrites in caned meats was reduced in several countries to the level needed to inhibit the growth of the bacterium responsible for botulism poisoning. The risk for the production of carcinogenic oxidized nitrogen compounds is high in the presence of readily nitrosylated drugs or pesticides. Readily nitrosylated drugs (Ethambutol, Dulcin, Piperazine, Aminopyrine, Methadone, Quinacrine) should be administered by a non-oral route (rectally or by injection) or formulated with enough ascorbate to block the gastric formation of nitrite compounds.

Many useful drugs are amines that can give rise to carcinogenic nitrosoamines. Here the only solution seems to be a risk/benefit evaluation with each drug, possibly with each patient. The question of whether to take cimetidine for dyspepsia, whether or not an ulcer is known to be present, or to take this drug indefinitely to prevent the recurrence of an ulcer requires careful thought and justification.

The levels of nitrosoamines to which man is exposed may be low at any given time, but the effect of one nitrosoamine given at different times is much than the simultaneous administration of different nitrosoamines, in which case the effect is additive. The nitrosoamines consumed in Scotch whiskey must be added to those in bacon, mushrooms, cigarette smoke, cosmetics, and to the nitrosoamines formed *in vivo*.

While the entire population is probably exposed to nitrosoamines in some form, everyone does not develop cancer. The more that is understood about the means by which nitrosoamines cause cancer, the more it is possible to detect high-risk individuals or groups, and to take protective measures. The best working hypothesis is that metabolism of nitrosoamines leads to the formation of alkylating species that react with DNA and cause mutation-like events, which disrupt certain control mechanisms in the cell.

The effect of components in a normal diet, for example alcohol, on metabolism of nitrosoamines is obviously of great importance in assessing the role of these compounds in human cancer. Also, the occurrence in cooked food of chemicals that increase the mutagenicity of nitroso compounds, causing previously inactive compounds to become mutagenic, is a cause for concern.

One in four of the population develops cancer and much of this is thought to be due to environmental agents that are, at present, mostly unidentified. The universal distribution of nitrosoamines and the variety of tumors they are known to induce in animals make it difficult to correlate nitrosoamine exposure with any particular human cancer. However, there is now epidemiological evidence that strongly suggests an association of N-nitroso compounds with esophageal, bladder, and stomach cancer. The fact that a variety of measures are now being taken to reduce exposure is a cause for congratulations to those engaged in this work.

9.5 OTHER HEPATOTOXIC COMPOUNDS

9.5.1 POLYHALOGENATED BIPHENYLS

Only in 1974 was the biological activity of these chemical pollutants discovered following a massive intoxication of cattle and other domestic animals near Battle Creek, Michigan. Approximately 20,000 animals and 1.5 million chickens had to be destroyed after being contaminated with polybrominated biphenyls (PBB) produced by a nearby factory.

Polyhalogenated biphenyls (PHB) are currently produced at the rate of 5,000 tons/year and are mainly used in the production of thermoplastic materials for the electronic industry. Their toxicity is magnified by their stability in the environment and the capacity of organisms to absorb and concentrate the compounds. Fish, for example, can concentrate PHBs 10,000 fold.

PHBs are mostly metabolized by biotransformation enzymes in the liver. These compounds are as potent inducers as 3-methylcolanthrene and phenobarbital. It was surprising to discover the inducing effect of PHBs is of very long duration, as much as 200 hours following a single administration. PHBs possess a mixed induction capacity, inducing biotransformation components that would require the action of both 3-methylcolanthrene and phenobarbital. PHBs significantly increase the hepatotoxicity of carbonitrachloride and bromo-benzene [89, 154].

The induction of hepatic microsomal enzymes resulting from a single injection of PBB indicates that the inducing properties change from being similar to phenobarbital initially to similar to methylcolanthrene. It has been postulated that, as occurs in phenobarbital administration, an epoxide intermediate is formed. Very little is actually known about the metabolism of this compound in man. It was demonstrated that, following injection of PBB, only modest increases of peroxides and cytochromes are found in the liver. But, a significant decrease in vitamin A and glutathione peroxidase are noticed. In animal studies, administration of 10 mg of vitamin A daily provides protection. It seems that PBBs may modify the toxicity of other, secondary, agents and alter the efficacy of certain therapeutic agents [89, 154].

9.5.2 AFLATOXINS

This is a group of mycotoxins possessing properties originally ascribed to an extract of the mold *Aspergillus flavus*, mostly found on peanuts (Arachis). Now, at least 40 molds (*Aspergillus, Penicillium, Phisopus*) are known to produce aflatoxins. Aflatoxins are present in most rotting food stuffs and can be present in food that does not physically seem spoiled.

The biosynthesis of aflatoxin is favored by:

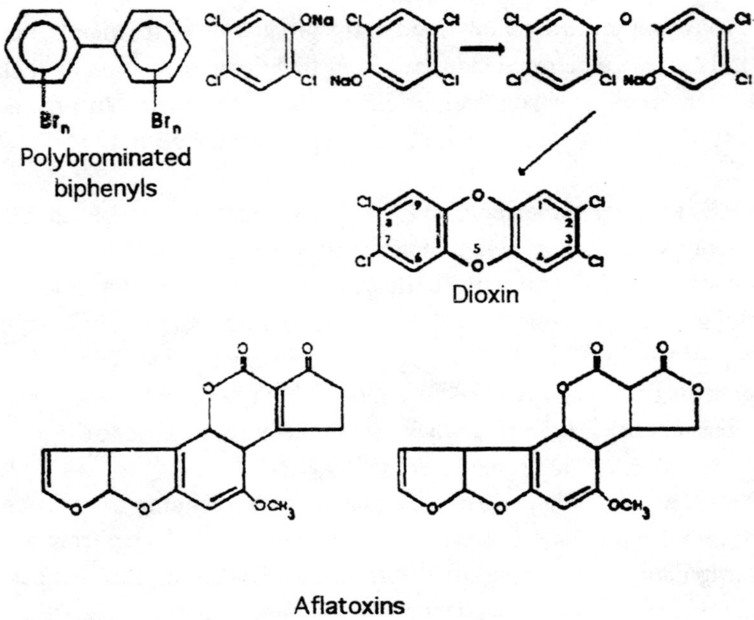

Figure 9.4. Other hepatotoxic chemicals present as chemical pollutants.

- Temperatures between 14° C and 42° C
- High humidity (over 80%)
- Slight acidity (pH 4 - 6)
- Low light and appreciable oxygenation

The toxicity of aflatoxin is among the strongest known. The LD_{50} is approximately 1 mg/kg body weight. Aflatoxins present some terrible biologic effects:

- Inhibit the growth of plants
- Inhibit the growth of many bacteria
- Strongly hepatocarcinogenic in both acute and chronic (10 μg/day) exposures

The carcinogenic property has been observed in children from Southeastern Asia following the ingestion of moldy peanuts. This carcinogenic effect occurs by the proliferation of the biliary channels. Aflatoxin strongly binds DNA at the N-7 of guanine and inhibits DNA polymerase. Chemically, aflatoxins are coumarin derivatives with variations due to different double bonds [158, 253, 263]. Derivatives of aflatoxins are given a letter and number identifier: aflatoxin B_1, B_2, G_1, G_2, etc. (Fig. 9.4).

Studies on aflatoxins are difficult and are not widely done due to their high toxicity. Glutathione seems to be the single natural compound that offers partial protection against aflatoxins. The high level of glutathione in liver (around 5 mM) seems to protect against low or moderate levels of exposure [40, 158].

9.5.3 HERBICIDES AND DIOXIN

There is abundant and convincing evidence that TCDD (dioxin or 2,3,7,8-tetrachlorodibenzo-*p*-dioxin [fig. 9.4] is an extremely potent carcinogen and general toxin in rodents, perhaps the most potent yet tested. Although a relation between risk of malignancy at several sites and exposure to dioxin has been observed in animals, the human carcinogenicity of TCDD remains controversial.

Massive uncontrolled human exposure has occurred in a couple of situations. During the Viet Nam war, herbicides were used on a large scale, as much as 50 kg/ha, building up an environmental concentration of 30 ppm. In 1968, explosions occurred in a herbicide plant in Derby, England and in Sevesso-Milano, Italy in 1967.

TCDD is actually an impurity generated during the manufacture of the herbicide 2,4,5-trichlorophenoxyacetic acid (2,4,5-T). Dioxin, or TCDD, is the strongest toxin ever known with an LD_{50} of 1 ng/kg (see comparisons in table 9.8).

This compound has a great ability to penetrate membranes. It acts on various levels, beginning with hepatic lesions, hemorrhage, bone problems, etc. The mutagenic and teratogenic effects are due to the protein-dioxin complex that penetrated further and covalently bind to DNA [77, 104]. As a result, severe modifications of the structure and function of genes and chromosomes are observed. As in the case of benzo-pyrene, a 7,8-diol-9, 10-epoxide is formed, possessing strong mutagenic and carcinogenic properties. Chronic human exposure to dioxin shows various clinical manifestations:

- Severe acne among workers at a chemical plant in Germany in 1954. This type of acne could also be produced in rabbits by dermal application of TCDD.
- TCDD was the causative agent of porphyria cutanea tarda reported in workers manufacturing 2,4,5-T.
- As a cause of Hodgkin's disease, nose and liver lesions, TCDD is subject to conflicting reports [81, 154].

Table 9.8
Comparative toxicity of selected poisons

Substance	Molecular weight	Minimum lethal dose (moles/kg body weight)
Botulism toxin A	9×10^5	3.3×10^{-17}
Tetanus toxin	1×10^5	1×10^{-15}
Diptheria toxin	7.2×10^4	4.2×10^{-12}
TCDD	322	3.1×10^{-9}
Saxitoxin	372	2.4×10^{-8}
Tetrodotoxin	319	2.5×10^{-7}
Bufotoxin	757	5.2×10^{-7}
Curare	696	7.2×10^{-7}
Strychnine	334	1.5×10^{-6}
Muscarine	210	5.2×10^{-6}
Diisopropylfluorophosphate	184	1.6×10^{-5}
Sodium cyanide	49	2.0×10^{-4}

It has been demonstrated that in some regions, including the Baltic Sea, fatty fish such as salmon and herring contain high levels of dibenzoxins and dibenzofurans. In spite of then finding these compounds in humans (13 pg/g fat), the clinical implications remain uncertain. Contaminated workers from a manufacturing plant in Missouri had a serum level of 325 pg/g of fat and Viet Nam veterans have 46 pg/g fat [77, 156, 288].

There are large species differences in susceptibility to dioxin. The dog is 1,000 times less sensitive than the guinea pig. The cause of death is uncertain in all cases. In the rat, rabbit, and mouse, hepatic necrosis seems to be the cause of death.

TCDD is only slowly metabolized in soil and its microbial degradation is limited. Dioxin is also a potent enzyme inducer for biotransformation components including UDP-glucuronyl transferase and glutathione transferase [81, 232].

Chronic exposure to dioxin produces a large range of symptoms, including acne, edema, infections due to a decrease in the activity of the immune system, and cancer. This suggests multiple routes of action or target systems. The resistance of organisms to low levels of dioxin also presents varied forms and, in the case of hepatotoxicity, glutathione seems to be essential.

9.5.4 HYPOLIPIDEMIC DRUGS AND PEROXISOMES

Peroxisomes are organelles discovered in bacteria (microbodies), plants, and the liver of animals by the Belgian scientist C. De Duve. Peroxisomes are rich in oxido-reductases, which produce keto acids by the reaction:

$$\text{Amino acids (hydroxyacids)} \xrightarrow{\textit{oxidase}} \alpha\text{-keto acids} + H_2O_2 \qquad (9)$$

In order to decompose the hydrogen peroxide produced in the above reaction, peroxisomes are also rich in catalase, which can mediate catalase (10) or peroxidase (11) type reactions:

$$2H_2O_2 \xrightarrow{\textit{catalase}} 2H_2O + O_2 \qquad (10)$$

$$H_2O_2 + \text{phenols} \xrightarrow{\textit{catalase}} 2H_2O + \text{Quinones} \qquad (11)$$

Lowrey and Gende [169] observed, in peroxisomes, high amounts of the enzymes involved in the β-oxidation of fatty acids, a process mainly occurring in the mitochondria.

Following the administration of hypolipidemic drugs, the number of peroxisomes in hepatocytes greatly increases. This appears to be a result similar to enzyme induction. This observation suggests that peroxisomes act as a control or regulator for the metabolic transformation of carbohydrates into lipids. This agrees with the presence of peroxisomes in adipose tissue and their quantitative variation as a function of physiologic activity. Following a lipid rich diet, the number of peroxisomes increase as does the β-oxidation of fatty acids. Populations fed a lipid-rich diet also may have an increased frequency of colon cancer [63, 232, 264].

It has been demonstrated that industrial plasticizers and hypolipidemic drugs (clofibrate: Nafenopin, Cyprofibrate) favor the induction of peroxisomes and increase the frequency of hepatocarcinomas in animals. Experimental studies [175, 215] showed a positive correlation between the development of cancer, number of peroxisomes, and the formation of lipid peroxides and lipofuscin pigments.

With the increase in the number of peroxisomes and β-oxidation of fatty acids, the amount of hydrogen peroxide will be increased due to reactions (9) and (12). Reaction (12) is catalyzed by acyl-SCoA dehydrogenase and is FAD dependent.

$$EnzFAD + R\text{-}CH_2\text{-}CH_2\text{-}CO\text{-}SCoA + H_2 \rightarrow EnzFADH_2 +$$
$$R\text{-}CH{=}CH\text{-}CO\text{-}SCoA + H_2O_2 \tag{12}$$

For each acetate removed from a fatty acid, 1 hydrogen peroxide will be produced. Due to this prolific generation of hydrogen peroxide, the capacity of catalase (reactions (10) and (11)) will be surpassed and the excess peroxide will function as a mutagen.

Goel and Sbarra [96] observed that after 6 months of chronic administration of cyprofibrate to rats, and increased formation of peroxides, conjugated dienes, and lipofuscin occurs. This event is concurrent with a decrease in glutathione and dependent enzymes. Generally, carcinogenesis will take place in organs where a massive proliferation of peroxisomes has occurred due to induction by hypolipidemic drugs, a fat-rich diet, exposure to plasticizers, etc.

The induction of peroxisomes by the above mentioned compounds seems to be a defensive response of the organism against an increased uptake of lipids. But a massive increase in peroxisomes is accompanied by an increased generation of hydrogen peroxide as a by-product of reactions (9) and (12). The increase in hydrogen peroxide may surpass the capacity of catalase and glutathione peroxidase to remove it, especially when glutathione decreases and inhibitory, toxic quinones and aldehydes accumulate (reaction 11) [165, 189, 224].

The carcinogenic and mutagenic activity of hydrogen peroxide is now strongly supported and is based on the formation of the OH$^\bullet$ radical and its attack on DNA or by intermediary peroxides.

The remarkable phenotypic variation seen in patients with adreno-leukodystrophy probably is a reflection of the effects of genetic, immunologic, and environmental factors. This rare disease is a consequence of a deletion in a gene involved in peroxisome function, and all patients have increased plasma concentrations of very long fatty acids. This indicates a generalized disorder of peroxisomal b-oxidation.

According to Ames [7], cooked, and especially fried, meats contain pyrolized amino acids, caramelized carbohydrates, and acrolein, which will induce an increased proliferation of hepatic peroxisomes. Both endoplasmic reticulum and peroxisomes are induced by a fat-rich diet and increased exposure to certain xenobiotics. Both metabolic and morphologic modifications of such a nature favor the activation of oxygen with the consequences already described. The impressive resistance of organisms to this challenge is due to nonspecific antioxidant defenses in which glutathione plays a leading role.

Part IV: Other Pollutants and Factors Affecting Exposure to Pollutants

CHAPTER 10

METALLIC POLLUTANTS

10.1 GENERAL COMMENTS

When discussing chemical pollution, most people think of compounds like SO_2, CO, NO_x, dioxin, etc. Metals can also constitute pollutants, and lead is the most common polluting metal. Large amounts were once released from the combustion of gasoline to which it was added in the form of tetraethyl lead in order to improve engine performance. It is calculated that the use of one tank of leaded gasoline released 32 grams of lead. Except for the case of methyl mercury, in which mass intoxications have occurred, most intoxications with metals go unnoticed. Consequently, the biologic effects are difficult to conclusively prove.

Life cannot exist without the participation of small amounts of metals (Fe, Cu, Zn, etc.), but in higher amounts many metals are toxic. An example is cobalt, required as a component of vitamin B_{12}, but when added as a stabilizer for beer foam (1 ppm) 20 people died [203].

In respect to the role and physiological concentrations in the organism, metals can be divided into several groups:

A. *Macrominerals*, such as Na, K, and Mg that are present in the body in large amounts and their physiological concentration is under homeostatic control.

Table 10.1
Concentration of some microminerals present in the human body and tap water.

Element	70 kg man	Daily requirement (μg%)	Plasma concentration (μg%)	Tapwater concentration (μg%)
Iron	3 - 5	12	80 - 130	0.30
Copper	0.1	2	85 - 98	1.00
Zinc	2 - 3	3 - 8	111 - 115	5.00
Manganese	0.02	3 - 9	20 - 30	0.05
Cobalt	0.002	0.075	3 - 4	0.01
Molybdenum	0.005	1 - 2	2 - 3	0.04
Chromium	0.006	0.1 - 0.4	2 - 6	0.05
Vanadium	0.003	1 - 2	1 - 2	0.05
Selenium	0.004	0.01	10 - 20	0.01
Silicon	0.02	?	100	0.05

B. *Microminerals*, such as Fe, Cu, Zn, Co, Ni, Cr, Mo, Mn, V, Se, I, Si, F, and Li. These are essential for normal function of the organism but are found in amounts that are less than 0.01% of dry weight.

C. *Other microminerals*, such as Ba, Br, Rb, Sr, As, Ge, Ti, Al, and Sn. Some of these may be essential.

D. *Toxic elements*, such as Pb, Hg, Cd, U, etc. Most of these are chemical pollutants.

Metal-containing drugs are, or have been, commonly used, adding to our exposure: silver (disinfectant), arsenic (Fowler's liquor for treating skin diseases), boron (disinfectant), barium (radiological examination), bismuth and aluminum (treatment of stomach ulcers and gastritis), mercury (for treating syphilis and as a disinfectant).

Metals are generally randomly and widely distributed in nature, making their assimilation by plants inevitable. A number of metals and their concentration in various sources are shown in table 10.1.

The great variation in metal concentrations in the body's tissues and plasma is only partially explained by their presence in the active sites of enzymes. The reactivity of metals and their presence in the environment obliged primitive forms of life to find ways of using or detoxifying them. Both procedures are based on the high binding capacity of proteins for metals.

This binding capacity can be reversible and temporary, as in the transport of metals, or irreversible with strong covalent binding. Therefore, new metaloprotein complexes were formed with new functions: enzymatic (cytochromes, catalase, etc.) or as a carrier (hemoglobin, transferrin, etc.).

Usually, when the binding of metals to amino acids or proteins is discussed, no toxicological point of view is given. The macrominerals easily fit into different structures and functions of organisms (enzymes, carriers, transmission of nerve impulses). For the microminerals (list B), evolution into usable compounds was not so straight forward. Most microminerals used by organisms are transition elements. These are characterized by the peculiar electron configuration in which the d and f atomic orbitals are only partly filled. That electronic structure confers great reactivity on these elements. In addition, because their ions possess an odd number of electrons, they favor the formation of free radicals and easily complex with a variety of ligands (amino acids, proteins, purines, pyrimidines).

A major feature of the transitional metals is the considerable decrease of toxicity that results from the formation of complexes. This explains why the macrominerals (A above: Na, K, Ca) can be found free in the ionized state in the blood and tissues, while the microminerals (list B) are mostly found stably bound. During evolution, microminerals from the $3d^2 4s$ group of transition elements were favored (Mn, Fe, Co, Ni, Cu, Zn). For this particular group, the ability to complex with other compounds and their stability depends on the electronic configuration of ions, atomic diameter and valence. These elements exist in several redox states, varying from zero (or negative) to +8. This variability allows their participation in a large number of complexes, thus increasing the possibility of forming stable complexes in different environments. The stability of complexes for ions in the $3d$ series increases from $Mg^{2+} < Mn^{2+} < Fe^{2+} < Co^{2+} < Ni^{2+} < Cu^{2+} < Zn^{2+}$ regardless of the ligand to which they bind.

The macrominerals have homeostatic mechanisms providing very efficient regulation of their concentration in the fluids of organisms. For the microminerals, the physiologic concentrations are maintained by the formation of complex, stable combinations.

Table 10.2
Nature and role of iron and copper combinations

Role	Iron			Copper		
	1	2	3	1	2	3
Transport	Transferrin	$2Fe^{3+}$	80,000	Ceruloplasmin	8 Cu	150,000
Storage	Ferritin	V	450,000	Hepatocuprein	V	35,000
Enzymes	Cytochrome oxidase	1 Fe^{2+}	230,000	Tyrosinease	Cu^{2+}	
				Dopamine hydroxylase	Cu^{2+}	
	Catalase	1 Fe^{3+}	250,000			
	Peroxidase	1 Fe^{3+}	40,000			
Metalo-proteins	Hemoglobin	Fe^{2+}	65,000	Monamine oxidase	4 Cu^{2+}	225,000
	Myoglobin	Fe^{2+}	18,000			

1=Common name, 2=Number of metal ions per molecule (V=varied number of atoms), 3=Molecular mass

As shown in table 10.2, there are a variety of combinations of Fe and Cu to be found in the body. Not all possible combinations are presented. Therefore, the formation of complex combinations of these essential metals may be regarded as a form of adaptation, necessary to decrease the toxicity presented by the free, ionized elements. Such an efficient solution was used for several metals, including Fe, Cu, Zn, Mn, Mo, etc. For ligands, organisms used amino acids (histidine) and heterocycles (imidasole or porphyrin). The many ways these metals have bound to ligands and the many functional molecules that result is the subject of a massive literature that will not be further discussed here [64, 105, 275].

One additional point needs to be examined a little closer; the binding of metals to nucleic acids. This binding may induce mutations or carcinogenesis. Generally, the binding of metals to nucleic acids can take place at the N-6 atom of purine or pyrimidine, or to the OH group of ribose or to the phosphate residue. With the binding of metals to nucleic acids, genetic and biosynthetic activities may be influenced or modified. The binding of metals to nucleic acids may be regarded as another example of the regulation of vital processes.

As shown in table 10.3, metals strongly contribute to the stability or scission of nucleic acids. *In vitro* experiments [64, 107] proved that the insertion of a metal into the structure of a nucleic acid or protein strongly modified the electron charge distribution and the tertiary structure. The nature and concentration of metals may induce reversible conformational modifications of nucleic acids. As seen in table 10.3, magnesium seems to be the preferred metal to provide stability to the DNA macromolecule. But, due

Table 10.3
The action of some metals toward nucleic acids

Nucleic acid and the affected process	Reaction or principle effect	Metals
DNA	Stabilization	Mg, Co, Ni
	Destabilization	Zn, Mn, Cd, Cu
RNA	Stabilization	Mg
	Destabilization	Zn, Ni, Cu, Pb, Cd
Catabolism DNA	DNAase activated	Mg, Mn, Co
Catabolism RNA	RNAase activated	Mn, Co, Ni, Zn
Replication DNA	DNA polymerase	Mg, Mn, Zn, Co
Transcription	RNA polymerase	Mg., Mn, Co
Transduction	Proteins, amino acids	Mg, Mn, Cu

to their closeness in atomic diameter and redox properties, Mg could be replaced by other metals in the same group, such as Zn or Mn, or in the unfortunate case Pb or Cd.

The replacement of some microminerals with others of similar atomic diameter and redox properties may explain the production of some genetic errors or mutations. The possibility of replacing the normal micromineral with another begins with their absorption through food and water. Some elements interfere or compete with others for digestion, absorption, and binding to transport and final complexes (table 10.4). This competition is modified depending on the form and combinations present in the food or water. Organic or inorganic molecules (phosphates, sulfates, citrate, oxalate, etc.) may increase or decrease the absorption of the minerals by the formation of more or less soluble complexes.

The formation of these complexes may explain the great capacity of organisms to adapt to increased concentrations of certain minerals in the environment, but also may account for the organism's concentration of some minerals. Some organisms possess the capacity to bind and concentrate metals to remarkable levels: some crustaceans and cephalopods are able to concentrate Si 13,000 fold; some mollusks concentrate Cu 4,000 fold, F 7,000 fold, and Pb 2 fold. Fish are able to concentrate phosphorous a million fold and collect Hg, Cd, Pb, and even radioactive elements such as I^{131} and Sr^{90}.

Table 10.4
Interference in absorbtion and distribution of microminerals

Element	Interferes with
Iron	Co, Cu, Mn, Ni
Zinc	Cu, Cr, Ca
Cadmium	Zn, Se, Mn, Cu
Copper	Mo, Mg, Zn
Selenium	As, Cu, Cd, Hg
Molybdenum	Cu
Aluminum	F
Manganese	Fe

The toxic elements, Pb and Cd, may be incorporated both by the formation of complexes and by the action of cellular ion pumps that are normally active towards Ca and Mg. The capacity to concentrate minerals is the result of the presence of a class of proteins, the metalloproteins, that are able to bind a variable number of metal atoms per molecule. Each molecule of ferritin can bind a thousand iron atoms. Subsequently, these metaloproteins are stored or eliminated. The storage of metaloproteins may take the form of granules of mixed structure about 1 - 10 μm in diameter. Such granules have been found in all existing organisms and are of various dimensions, from microscopic to macroscopic. Because nature keeps any useful thing, some of these granules became involved with the vibrating cilia of the inner ear, providing equilibrium and the conversion of sound waves into electrical impulses.

10.2 LEAD

As mentioned earlier, lead is the polluting metal that attracted the attention of public opinion regarding its biological consequences. This concern dates as far back as the Roman Empire, when lead was widely used for the transport of water (pipes), an additive to wine, food, etc.

Table 10.5
The environmental concentration of lead.

Medium	Unit	Area		
		Rural	Urban	Near smelters
Soil	$\mu g/g$	5 - 50	10 - 50	20 - 2,000
Air	$\mu g/m^2/day$	<0.3	0.3 - 3	300 - 20,000
Dust, fallout	$\mu g/g$	20 - 80	50 - 300	500 - 10,000
Street dust	$\mu g/g$	50 - 200	600 - 3,000	1,000 - 15,000
Drinking water	$\mu g/L$	1 - 10	5 - 30	10 - 1,000

Table 10.5 presents some examples of the environmental concentrations of lead. The exact values will vary with country and geographic area. In the presence of atmospheric oxygen, lead becomes susceptible to attack by acids, including very weak acids like carbonic acid or the acid in rain. Therefore, the presence of dissolved lead in fresh water is a well known phenomenon. But lead solubility is considerably reduced by the presence of small amounts of carbonate or silicate in the water, and decreases further with water hardness [115, 174, 197, 203].

In 1985, the total consumption of refined lead was about 6 million tons, world wide. About 40% of this was used for the production of lead-acid batteries, while tetraethyl lead accounted for 10%. Alcoholic beverages, especially wine may contain substantial amounts of lead: mean values of 50 - 100 $\mu g/L$. High lead levels have been found in illegally distilled liquors (vodka and whiskey) due to the presence of lead-soldered joints in the distilling apparatuses. Wines and spirits stored in lead crystal decanters and glasses for a long time contained lead at concentrations up to 22,000 $\mu g/L$ [99] as presented in table 10.6. Data in the right half of table 10.6 was obtained by storage in crystal in use in North America prior to the introduction of lead crystal.

It is generally agreed that the measurement of lead in the blood is an indicator of absorbed lead. On the average, blood lead levels in the general population, free of occupational exposure, are in the range of 5 - 20 $\mu g/dl$. This level is strongly influenced by a number of demographic, social, and geographic factors in addition to individual variation.

Table 10.6
Lead concentration in spirits stored in lead crystal decanters (left) and non-lead glass (right). After Graziano [99].

Beverage	Approx. storage time	Lead conc. (μg/L)	Beverage	Approx. storage time	Lead conc. (μg/L)
Brandy	5 years	7,746	Gin	1 year	13
Madeira	5 years	1,402	Brandy	1 year	68
Whiskey	> 3 years	2,687	Bourbon	18 months	17
Port	1 year	203	Tequila	18 months	300
Armagnac	1 year	203	Vodka	1 year	11
Gin	5 years	8,390			

Absorbed lead enters the blood from where it is distributed to various organs and tissues. In blood, lead is mainly bound to the hemoglobin within the erythrocytes and to the albumin of the plasma.

Lead is removed from the body at a very slow rate. The biliary route is of greater importance in the excretion of lead than is the urinary route. Lead is transported from the plasma to the bile against a concentration gradient, as demonstrated by Mahaffey [174]. The large bile/plasma concentration ratio supports the concept that lead is excreted into bile by an active transport process. This is further supported by the high affinity of the liver for binding lead. This high ratio (34 for a lead intake of 0.1 mg/kg) decreases dramatically with increased concentration (0.6 for a lead intake of 10 mg/kg). This decreased ratio at higher concentrations is further support for the presence of an active process.

Lead alters several proteins and processes:

a) Inhibits δ-aminolevulinic acid dehydratase (ALAD), which is important in the synthesis of heme. This enzyme is completely inhibited when blood lead levels reach 70 - 80 µg/dl. Urinary ALA is frequently used as an indicator of lead exposure.

b) Inhibits ferrochelatase, the enzyme that inserts iron into the porphyrine ring to convert it to heme. When this enzyme is inhibited, the porphyrine binds zinc and accumulates in the blood.

c) Anemia is a well known effect of lead intoxication, due to altered hemoglobin and reduced hemoglobin synthesis. The threshold for anemia is 50 µg/dl. Children are most sensitive.

d)   Heavy exposure to lead may impair renal function due to renal tubule damage. Creatinine clearance is decreased because lead deposits in the proximal tubules [197, 279].

e)   Neuropsychological abnormalities due to alterations of nerve fibers and encephalopathy. Restlessness, irritability, headache, muscle tremor, hallucinations, loss of memory, and loss of ability to concentrate are common manifestations of lead poisoning. Again, children are the most sensitive with a blood threshold of 50 µg/dl. For a more detailed discussion see Baghurst [16].

f)   Cardiotoxic effects. Hypertension appears at a blood concentration of 40 µg/dl.

g)   Chromosomal aberrations and alterations in lymphatic function have been noticed.

In adults, the most important clinical manifestations following lead poisoning are (in descending order): abdominal pain, constipation, vomiting, muscle weakness, numbness, and psychological symptoms. Mild contamination produces tiredness, lethargy, constipation, abdominal discomfort, and sleep disturbances. The presence of a blue line in the gums (Burton's line) and a metallic taste in the mouth are useful indicators of increased lead absorption.

Lead, as an element, has no useful purpose for the living organism. Its altering effects are based on its binding to SH, NH_2, carboxy, and imidazol residues in proteins. This activity explains the vast array of clinical manifestations. Alterations to neuropsychological functions seem to be the result of alterations in dopaminergic receptors [16, 73]. Lipofuscin stores are the result of lipid peroxidation and is especially noted in the nigro-striatal region of the brain. Destruction of the adjacent dopamine producing substantia nigra cells causes Parkinson's Disease.

The organism has poor defenses against this aggressive element. For treatment the use of chelators, such as EDTA or DTPA, is only effective immediately following an acute exposure. The absorption of lead is slightly decreased by ascorbate, phosphates, and citrates. The glutathione concentration is also affected in lead poisoning as it attempts to protect against the formation of peroxides. Alcoholic beverages increase the absorption and toxicity of lead.

10.3 MERCURY

In many ways, mercury presents something of a paradox. It is a metal, but is liquid at room temperature. Its medical implications also present a paradox. Mercury has been administered orally since ancient time in doses as large as 100 - 500 grams (for intestinal occlusion), its vapors are toxic and have a high vapor pressure (10 mg/m^3 at 22° C). HgCl (calomel) is insoluble and has been administered orally as a cathartic agent, while $HgCl_2$ (sublimate) is a strong poison having a lethal dose of 0.20 g. The toxic effects of pollution by mercury come mostly from its organic salts (methyl and alchyl-mercurates), but organic mercury compounds are used as diuretics. In spite of its uses, mercury remains a major chemical pollutant.

Mercury is widely used in the chemical and electronic industries, meteorology instrumentation, and agriculture (pesticide and fungicide). The ecological risks come from industrial contamination of the environment or the introduction of methylmercury insecticide into foods and cereals.

Recent reports [167, 225, 238] indicate mercury is released from dental amalgam (possible 10 µg/day absorption) and mercury vapor is released from latex paint (median concentration of 10 nmol of Hg per m^3 of room air). For most people, the daily intake of mercury ranges from 1 - 30 µg. The current allowed limit for mercury in air is 0.05 mg/m^3. Total blood mercury ranges from 6 µg/L in people with low fish consumption to 50 µg/L for people eating moderate amounts of tuna and up to 200 µg/L after eating large amount of predatory marine fish. About 95% of the mercury found in whole blood is located in the red blood cells, mostly as methylmercury. Hair concentrations correlate well with blood concentrations, but urine levels are not a good index to the body burden of methylmercury. Metallic mercury vapor is about 75% absorbed by the lungs during inhalation. Urine concentrations are the best indicator of blood levels following exposure to inorganic or metallic mercury. In normal individuals, urine levels are around 10 µg/L, but large individual variation exists.

In the late 1940s, pediatricians realized the connection between mercury and acrodynia (an allergic reaction to mercury in young children, now a rare disease) in children, in which the sympathetic nervous system seems to be a major target. Strangely, in mass outbreaks of methylmercury poisoning in

Japan and Iraq, no cases of acrodynia were reported. Because of this, it has been suggested that acrodynia is the result of exposure to metallic mercury.

Another characteristic of mercury is the biotransformation process that takes place in nature. Irrespective of its form (metallic, inorganic, or organic), mercury is converted by anaerobic bacteria at the bottom of lakes and ponds to methylmercury (CH_3Hg^+) and dimethylmercury (($CH_3)_2Hg$). The latter, more volatile, compound passes into the atmosphere (1 ppb). As can be seen from figure 10.1, Hg is converted to Hg^{2+} and then to CH_3Hg^+. Microorganisms methylate mercury according to the following enzymatic reaction:

$$Hg^{2+} \xrightarrow{\text{methylating enz.}} CH_3Hg^+ \to (CH_3)_2Hg$$

There is a secondary route involving vitamin B_{12}, presented in figure 10.1B that takes place in the intestine with the help of the resident microbial flora. Mercury is also methylated in the liver, but in that organ nonenzymatic demethylation also occurs, through the action of the SH groups of glutathione and cysteine. Mercury has a high capacity for binding to, or precipitating proteins, especially those with SH groups. Due to this high binding capacity, mercury binds to various proteins and accumulates in various organs, especially the kidney. Mercury is slowly excreted from the body, and the rate at which it is excreted depends on the chemical form administered. Methylmercury, widely found in the environment, is excreted more slowly than other mercurial forms, having a half-life of 70 days. The fecal route of excretion seems to be more important than the urine for the excretion of mercury compounds.

The biochemical consequences of mercury binding to proteins can be grouped as follows:

a. Inhibition of adenylcyclase and, by implication, the metabolic regulation of cAMP.

b. Mercury salts bind to plasma proteins and erythrocytes, so penetration into the tissue is slow. Elimination by the kidneys is achieved through complexing with thiols, such as cysteine. The normal urinary values range from 0 - 40 µg/L. Due to the high affinity of mercury for free protein SH

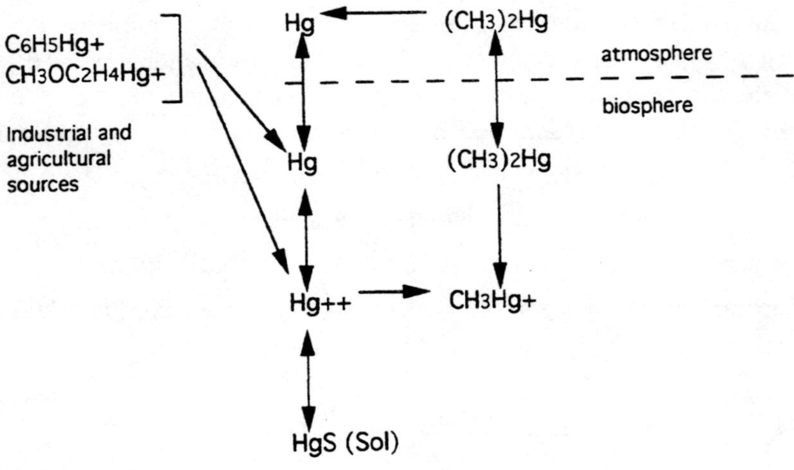

A

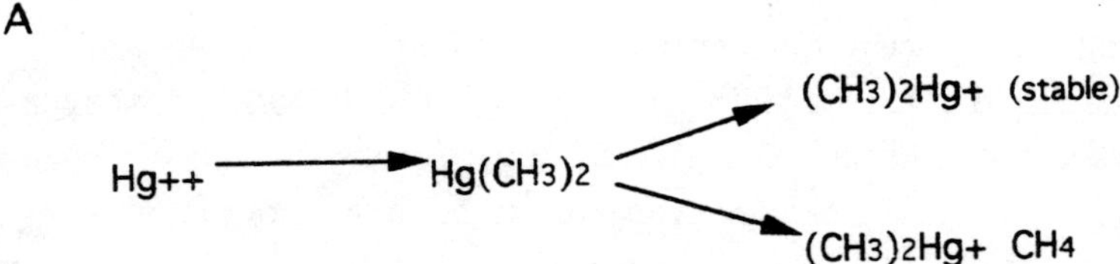

B

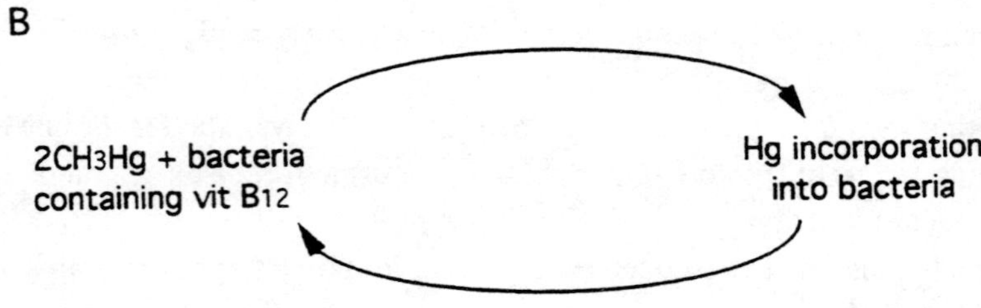

Figure 10.1. Methylation of mercury in nature. After Petering [222].

groups, a relatively small amount of mercury is enough to upset the metabolic process that depend on enzymes containing SH groups in their active center.

c. Mercury can be transported to the kidneys bound to the SH groups of hemoglobin.

d. Binding of mercury to nucleic acids may cause mutations, metabolic deficiencies, congenital retardation, and perturbation of mitochondrial or lysosomal functions [225, 238].

e. Methylmercury and mercury vapor (entering by alveolar blood) reach the brain and attack the nerve cell's membrane plasminogens, releasing acids and aldehydes. This results in behavioral changes (apathy with episodes of irritability and depression) called erethism that are characteristic of the early stage of chronic intoxication. This is followed by advanced stage symptoms of trembling fingers, eyelids, and lips, blurred vision, loss of coordination (ataxia), and difficulty swallowing (dysphagia).

f. Petering and Tepper [222], in their studies on volunteers using Hg^{203}-labeled methylmercury, found that the half-life is 70 - 80 days. The brain stores 10% and the liver about 50% of the administered methylmercury. It seems to be primarily excreted in the feces through the bile. Mercury has a high affinity for bile. Bile's mercury concentration exceeds that of the plasma. Experiments carried out on hepatocytes incubated *in vitro* with $HgCl_2$ (200 µM) showed toxicity due to a 60% decrease in intracellular K+ and an elevation of glutamate-pyruvate transaminase (40%) and peroxidation (50%).

g. A highly controversial issue is the competitive and possibly protective role of selenium in chronic mercury intoxication. This is a complex question currently being studied biochemically, physiologically, and geochemically. Mercury binding to proteins changes their tertiary structure, creating new centers of great affinity. The protective role of selenium differs depending on the salt used, decreasing in effectiveness in the order selenomethionine > selenium extracted from tuna > selenite. The protective role of selenium has been evaluated in relation to specific criteria: rate of growth, morbidity, nervous disorders. The presence of mercury (41 ppm) in the brain of rats following a 9-week administration of a diet containing methylmercury (20 ppm) and selenite (5 ppm) failed to produce lesions or neurologic symptoms. Normally, selenium penetrates the blood-brain barrier with difficulty, but in the presence of mercury, the rate of accumulation is higher. The high mercury concentrations found in fish, especially ocean fish, is always associated with a high selenium store, which may represent a protective adaptation [167, 225].

Mercury may act differently depending on its concentration by stimulating Ca-ATPase and liver dihydrofolate reductase. Conversely, cholesterol synthesis (at the mevalonate-squalene stage), hemoglobin biosynthesis, and the electron transport chain in the mitochondria are inhibited even at concentrations of 10^{-6} M. Cellular respiration is dependent on the form in which mercury is present and its solubility in lipids.

Mercury has been incriminated as being neurotoxic, teratogenic, embryotoxic, and carcinogenic. With chronic intoxication, the symptoms are cerebellar disturbances (95%), sensory impairment (65%), visual problems (60%), speech disturbance (72%), pyramidal lesions (38%), and encephalopathy (32%). Methyl mercury also affects the immune system, altering antibody production.

Mercury intoxication is treated with chelating agents that act through SH groups, such as BAL (dimercaptopropanol), DL-penicilamine, β-β-dimethylcysteine; "catatoxic steroids," such as spironolactone (protect from nephrotoxic effects), glutamic acid in high concentrations (2 g/kg), and selenium compounds, although the latter have a problem of toxicity.

The weekly tolerated exposure is 0.3 mg Hg, giving a red blood cell concentration of 0.5 µg/dl and twice that in the plasma.

10.4 ALUMINUM

Until recently aluminum was not considered a toxic metal, however there is now evidence that the accumulation of aluminum in hemodialysis patients may cause encephalopathy and other serious complications. Aluminum has also been implicated in the pathogenesis of other neurological conditions. Consequently, following an increasing number of publications, there are claims that aluminum is a factor in causing dementia.

Aluminum is generally found as the relatively insoluble aluminosilicate, which comprises 8% of the earth's crust. The daily intake in man is approximately 21 mg by ingestion and 10 µg by inhalation. Aluminum, unlike most of the transition metals, has a fixed valence of +3 in biological systems, so is unable to participate in redox cycle reactions. Never-the-less, there is growing evidence to suggest that aluminum can act synergistically with iron ions to increase free radical dependent oxidation of polyunsaturated fatty acids, to increase permeability of the blood-brain barrier, to inhibit hemoglobin synthesis, and modify phagocytic activity [84].

Exposure to aluminum comes from the wide utilization of this metal in goods including cooking utensils, dishes, tooth paste, food wrap, drugs, and dyes. These uses result in an exposure that exceeds the exposure to iron. The possibility that aluminum might be toxic passed unnoticed because of its insolubility and hence low absorption in most forms. Aluminum enters the body by digestion and through breathing. Drinking water standards limit it to 50 μg/L (50 ppb).

As presented in figure 10.2, this metal is excreted by the kidney, but undergoes a process of concentration in the proximal tubule, followed by precipitation in lysosomes through the action of acid phosphatase. When there is significant absorption, aluminum may accumulate in various organs including the liver, brain (possibly resulting in early senile dementia), heart, bones, and bone marrow. Evidence of this has been obtained through the use of new, sensitive instrumentation such as the Casting probe (an X-ray spectrometer associated with the electron microscope) or analytic ionic microscopy [206]. Atmospheric aluminum and aluminum deposited in organs takes the form of lysosomal microcrystals of aluminum phosphate. What surprised investigators was the great ability of human body cells to extract aluminum from the blood, where the concentration may be only 10^{-7} - 10^{-8} g/L, and concentrate it 100,000 times in the lysosomes. X-ray spectra always show the association of aluminum with phosphorous. In the lysosome there is an appreciable amount of acid phosphatase that releases phosphate from organic compounds (breaks the ester bond). Due to the affinity of Al and PO_4, insoluble aluminum phosphate is formed and deposited. The process of concentration by acid phosphatase is beneficial to the body, particularly in the kidney where the nephrons divert the phosphates into the urine, eliminating them (figure 10.2). But at the same time, this property allows the storage of aluminum in various other organs that are rich in lysosomes: brain, liver, heart. Thus, there have been reports of lethal encephalopathy in patients on renal dialysis. Paradoxically, in the liver, where the concentration of lysosomes is the greatest and there are relatively large aluminum deposits, hepatic function is little affected until very high concentrations are achieved. The explanation for this lies in the capacity of hepatocytes to divide and

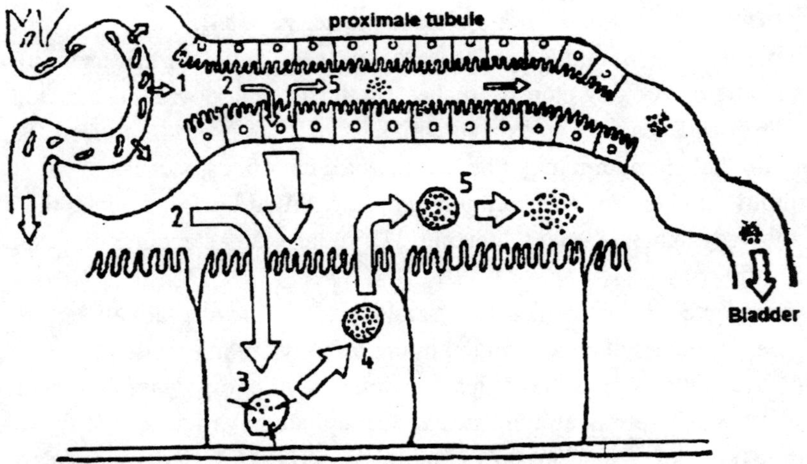

Figure 10.2. Renal excretion of aluminum. Aluminum is filtered from the blood through the glomeruls (1), reabsorbed into the proximal tubule (2) and precipitated in the lysosomes (3) in the form of phosphates (4) and excreted (5).

regenerate, a property the brain and heart cells do not have.

Aluminum-induce anemia occurs due to its storage in macrophages. These phagocytic cells play an essential role in the metabolism of iron. Iron, in the form of ferritin is stored in the siderosomes of macrophages and released when needed for hemoglobin synthesis. Aluminum upsets this iron storage.

Decalcification of bone with the occurrence of spontaneous fractures (especially in uremic patients) is due to the action of alkaline phosphatase and the competition between normal hydroxyapatite deposition and the formation of aluminum phosphate deposits. The result is a perturbation of the normal formation of the bone's hydroxyapatite microcrystaline network. Further studies are needed to improve our knowledge of the involvement of aluminum in human pathology.

The presence of increased amounts of aluminum in the brain of patients with Alzheimer's disease (AD) suggested a possible involvement of this metal in the etiology of this disease [84]. Aluminum is known to cause dialysis dementia (DD). This results in seizures, progressive cognitive changes, and results in death within a year. The high aluminum concentrations found in DD do not cause the amyloid plaques and neurofibrosis found in AD. Although DD can be fatal, it can be treated with some success with chelating agents. In DD, the brain is damaged by high doses of aluminum that results from kidney failure. In AD, lengthy exposure to low doses of the metal may cause a slow death of brain tissue. Although DD may be reversed by dietary restrictions and chelation, these only slow the progression of AD [188, 306]. Epidemiological studies made in South Carolina (1980), Norway (1986), England (1989), France (1990), Switzerland (1991), and Canada (1991) provided controversial results [291]. These studies suggest that normal drinking water may be a risk factor for AD.

As aluminum hydroxide is widely used as an adjuvant in the preparation of diphtheria, tetanus and pertussis vaccines, this may present another potential source of exposure. In England, the aluminum content is restricted to 1.25 mg per dose of vaccine. Experimental evidence shows an accumulation of aluminum in the brains of mice that were immunized with aluminum adjuvinated vaccines [206, 234].

Fortunately, most of the aluminum ingested (90% of exposure is by food) is not absorbed from the intestine. The absorption of aluminum is strongly dependent on the specific salt; citrate binding is particularly effective in increasing absorption.

10.5 CADMIUM

Unlike the metals presented earlier, cadmium appears quite unexpectedly on the list of chemical pollutants. The usefulness of Cd for the body is still unclear, but it is widely used in industry as an anticorrosive, in alkaline batteries, bearings, printing, plastics, treating disorders of the mouth, nuclear moderators, etc.

Pollution of the environment with cadmium occurs through burning natural fuels and by processing and using cadmium containing materials. This low but continuous exposure can be increased through foods (organ meat, cereals, 0.06 μg/Kg), drinking water (limit 1.10 μg/L), and smoking (2 - 4 μg per pack).

The body is not equipped with a homeostatic mechanism to regulate the cadmium level in tissues. The total amount of cadmium in the body of people not seriously exposed to it is about 30 mg. Cadmium concentrates in the liver, kidneys, testes, blood (0.6 μg/dl on average), and urine (5 - 25 μg/L). In workers exposed to cadmium-rich materials, the level in the urine is 10 - 100 times greater [135]. It is possible that the existence of large cadmium deposits in the liver and kidney may be due to the same mechanism that concentrates aluminum. For the general population, the accumulation by digestion and respiration is 10 - 40 μg/day. Excretion is very slow, proceeding mostly through the bile with the renal route only becoming significant in the presence of heavy exposures. Renal lesions occur at about 200 mg/kg tissue.

As noted in the introduction to this chapter, cadmium interferes with the adsorption of other microminerals, especially Zn, Ca, and Cu. Cadmium toxicity is due to its concentration in the liver and kidneys, coupled with its very low excretion rate (half-life of 10 - 20 years). It binds to proteins and exhibits Cd-Zn antagonism, replacing zinc in the structure of some enzymes (carbonic anhydrase, peptidase).

Signs of chronic intoxication occur at relatively low concentrations, varying from one person to another in the range of 30 - 50 mg. The signs are nausea, vomiting, cramps, diarrhea, lack of appetite, insomnia, etc. Cadmium was included on the list of polluting materials after industrial waste contamination with of the Jintsu River in Japan in 1968 caused "itai-itai" disease among those who drank the water. When administered to rats, cadmium may induce hypertension and increased renal sodium readsorption. The urine of hypertensive patients contains 10 - 50 times more cadmium than normal, and the renal level is also higher.

The testes retain cadmium. In rats (30 μg/kg) this results in hemorrhagic necrosis and irreversible involution. Because zinc is required for spermatogenesis, its replacement with cadmium may lead to disfunction or sterility.

In experimental chronic intoxication of rats (0.1 - 0.5 mg Cd/kg) for 5 weeks (15 doses), Klimczak and Kolakowsky [144] found impairment of the hepatic biotransformation system, increased peroxidation of lipids, and induction of hemoxygenase. These effects may be related through the mechanism described in chapter 4, in which either biotransformation enzymes or bilirubin leads to an increased peroxidation and deterioration of cellular and subcellular membranes. At a level of 200 µg Cd/g hepatic tissue, there occurs structural and functional alteration due to biochemical lesions (table 10.7).

Table 10.7
Effects on biochemical parameters in rat liver following chronic exposure to cadmium. After Klimczac and Kolalowsli [144].

Group	Cd (μg/g)	APH	ALAS	Oxygenase	Lipid peroxide
Control	0.40±0.03	10.0±0.5	7.6±0.4	63.3±2.7	4.01±0.26
Cd (1.5 mg/kg)	10.6±1.36*	10.6±0.8	6.4±0.2*	60.6±1.2	4.9±0.6*
Cd (7.0 mg/kg)	95.3±8.65*	7.4±0.5*	8.3±0.5	101.3±2.1*	5.6±0.5*

APH=Anilin p-hydroxylase. Values are expressed as nmol p-aminophenol/mg protein.
ALAS=Aminolevulinic synthetase. Values are expressed as nmol ALA/g.
Oxygenase=hemoxygenase. Values are expressed as nmol bilirubin/g tissue/30 minutes.
Lipif peroxide=μmol malondialdehyde / L

The kidney seems to be the most critical organ affected by chronic exposure to cadmium, although this metal also causes lung damage, osteoporosis, and osteomalacia. The first detectable nephrotoxic effect of Cd is an increased excretion of protein in urine, which cannot be detected by conventional hospital tests. The first group of uncommon proteins excreted in urine have a varied molecular mass and include aminopeptidase and N-acetyl glucosaminidase. Increased excretion of these proteins indicates increased turnover of renal tubular tissue [132]. The second group of proteins includes albumin and transferrin, indicating glomerular damage and an alteration of tubular reabsorption.

Cadmium uptake and absorption may be influenced by the presence of zinc, iron, calcium, and copper. Other factors also strongly influence cadmium absorption, such as protein consumption (possibly as a source of glutathione) and deficiencies in vitamins C, D, pyridoxine, and thiamin. Hypertension and diabetes are known to increase the risk of kidney damage [102, 131].

10.6 DEFENSES AGAINST METAL INTOXICATION; METALLOTHIONEIN AND STRESS PROTEINS

Mammalian MTs (metallothionein) consists of a single polypeptide chain, generally of 61 amino acids with a strong β-turn conformation. MTs have long lasting lives, 15 years for Cd-MT. Their induction takes place in erythroblasts and reticulocytes, but they disappear quickly from the blood. The induction, through transcription of specific genes, by metal ions may present a double edged sword: it provides the cell with MTs that act as a sink for toxic metals, but it may also cause the activation of oncogenes and growth factors that promote tumor growth. MT destruction occurs in lysosomes.

The best studied of these is the induction of metallothionein that occurs in animals, plants, and microorganisms. Metallothionein is a small protein consisting of 60 amino acids of which about one-third are cysteines. The protein has a particularly high metal-binding capacity. This protein is induced in mammalian systems by many metal ions: Cd(II), Zn(II), Cu(II), Hg(II), Ag(I), by glucocorticoid hormones, and by growth factors [238].

Class 1 MTs can be divided into two subgroups: MT-1 and MT-2, possessing two or three negative charges at neutral pH. Presently, the primary structures of 25 MTs are known completely or in part. The position of the 20 cysteine residues in the polypeptide are highly conserved (figure 10.3). Their arrangement in the Cys-X-Cys sequence renders the protein a potent metal chelator. In fact, all the cysteines are involved in the binding of bivalent metal ions, giving rise to a cysteine/metal ratio of about 3. This unusual stoichiometry indicates that the metal complexes are part of clusters in which neighboring metals are linked by bridging thiolate ligands.

A most striking effect is the reduction in cadmium toxicity for animals that have been pretreated by low doses of cadmium or zinc. This protective effect, however, is not due to a decreased body burden of cadmium. Administration of low doses of dietary cadmium in experimental animals caused an accumulation of cadmium, zinc, copper, mercury, and silver in the form of tissue metallothionein. The binding of metals to thionein may have adverse effects when patients or animals receive large oral doses of zinc. If they synthesize excessive thionein they may suffer from copper depletion because so much is bound to this protein. This type of interaction may be a

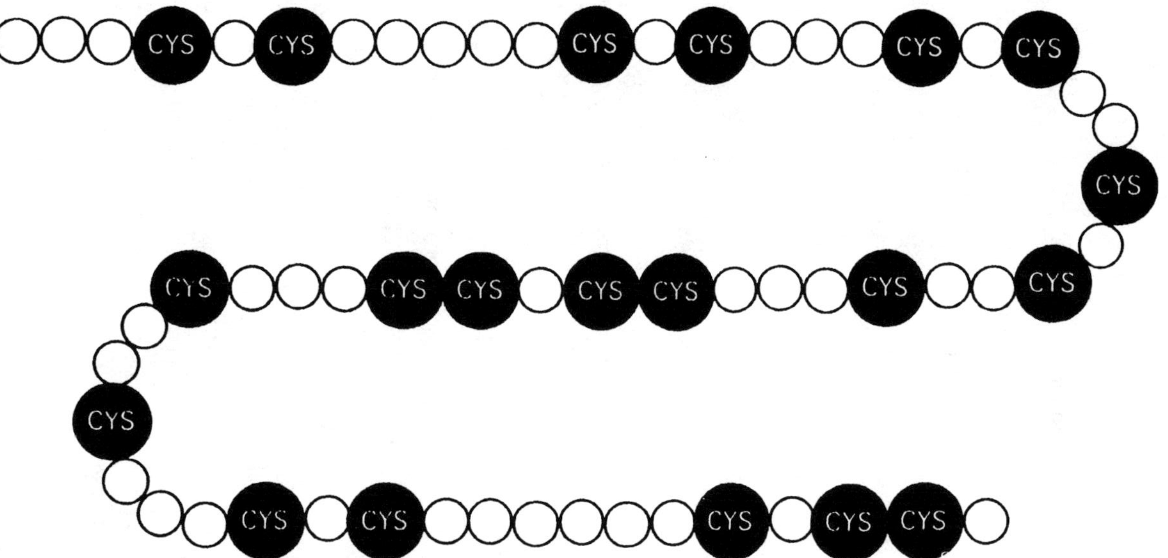

Figure 10.3. Schematic amino acid sequence of mammalian metallothionein. The small circles represent amino acid residues other than Cys. The diamond at one end indicates the amino terminal acetyl group.

Table 10.8
Inductive factors towards metallothionein synthesis in cultured cells or <u>in vivo</u>

Metal ions: Cd, Zn, Cu, Hg, Ag, Co, Ni, Bi, Au	Streptozotin
Glucocorticoids	Isopropanol
Protesterone	Ethanol
Estrogen	Eyhionine
Glucagon	Alkylating agents
Catecholamines	Chloroform
Interlukin I	Carbon tetrachloride
Interferon	Starvation
Butyrate	Infection
Retinoate	Inflammation
Phobol esters	Laparatomy
Endotoxin	Physical stress
Carrageenan	X-ray irradiation
Dextran	High oxygen tension

major mechanism of zinc toxicity.

In addition to thioneins, stress proteins are induced by certain metal compounds: Cu(II), Cd(II), and As(III). Stress proteins are better known as being produced in response to heat shock, being produced by short-term treatment of animal, plant, or yeast cells with an elevated temperature (about 42° C). In animal cell cultures, the induction of stress proteins by moderate concentrations of Cu(II), Cd(II), and As(III) induced a limited resistance to these metals. The mechanism of induction of protective proteins by metal compounds is still in debate. The finding that both metals and glucocorticoid hormones stimulate the same human metallothionein gene suggests a connection between the inducing metal ions and the transcriptional activation factors [86]. MT gene promoters have been identified and fused to a variety of other genes as a means to regulate gene expression by the stimuli presented in table 10.8

CHAPTER 11

MODIFICATION OF EFFECTS, TERATOGENIC EFFECTS, AND DETECTION AND CLEANUP OF ENVIRONMENTAL POLLUTANTS

11.1 MODIFICATION OF EFFECTS OF EXPOSURE

Chronic exposure to chemical pollutants may not result in any significant effect on people. Conversely, chronic exposure may alter the person's resistance, favoring the appearance of so-called degenerative diseases and accelerating aging. How the balance between the protective systems of the body and the injurious effects of the pollutant will be tilted depends on a range of environmental factors, chemical and physical form of the pollutant, and the body's resistance.

11.1 CLASSIFICATION OF MODIFYING FACTORS

The study of the influence of factors that modify biological processes has important implications for long-term drug therapy or chronic occupational exposure to pollutants. However, even when the chemical structure of the pollutant and the nature of the exposure are taken into account, it is frequently not possible to accurately extrapolate to the final effect.

11.1.1 FACTORS DEPENDING ON CHEMICAL STRUCTURE

The chemical structure of polluting substances is a very important consideration. The molecule may have aromatic rings, toxic groups ($=C=O$, $-S=O$, NO_2, $-N=C=$), or lipid solubility. The last characteristic assures effective penetration. Depending on the structure, biotransformation may neutralize of increase the toxicity of the molecule.

If the chemical pollutant can exist in several physical states, the gaseous form is generally the most toxic because of its easy penetration into the body by the lungs. The penetration of dust, aerosols, and powders may carry other chemicals along with them.

The most important factor is the *solubility* of the chemical pollutant. To penetrate cell membranes the chemical must be partially soluble in both water and lipids, giving the molecule a moderate lipid/water partition coefficient [63, 287]. For molecules that are more water soluble (e.g., ethyl alcohol) penetration depends on saturating the transport mechanism and a slow rate of biotransformation, allowing toxicity to develop through accumulation of the compound. If, on the other hand, the water solubility is low (organic compounds), both saturation and elimination take place rapidly, especially in the case of acute exposures as may occur due to occupation.

The *concentration or dose* is a very important factor for both chronic and acute or occupational exposures. Every country keeps official tables of maximal allowed concentrations (MAC) of chemical exposure (table 11.1). These standards are generally set to allow no more than a moderate exposure that will not produce clinical manifestations in workers who are continually exposed. People who are hypersensitive to the chemical will still be affected.

The values for MAC differ significantly between countries. Thus, for lead it varies from 0.01 mg/m^3 in Russia to 0.20 mg/m^3 in the USA and Germany. Similar differences can be found for radioactive fallout and metal contaminated food. This wide range is both an effect and an illustration of the wide variation in the resistance of individuals to chemical pollutants. Generally these limits are based on chronic exposure.

Table 11.1
Maximum allowed concentrations according to the Eastern European
Standards, 1980.

Compound	mg/m^3/24 hours	mg/m^3/30 minutes
Hydrochloric acid	0.1	0.3
Sulfuric acid	0.1	0.3
Ammonia	0.1	0.3
Aniline	0.02	0.05
Gasoline	2.0	6.0
Nitrogen dioxide	0.1	0.3
Sulfur dioxide	0.25	0.75
Chlorine	0.1	0.3
Phenols	0.03	0.1
Formaldehyde	0.01	0.1
Hydrogen sulfide	0.01	0.15
Mercury	0.001	0.003
Carbon monoxide	2.0	6.0
Lead	0.001	—
Oxidants	0.03	0.1
Dust	0.15	0.15

The relationship illustrated in figure 11.1 holds for chemical pollutants and radioactive fallout. The most interesting point to be made is related to the modification of the slope when the cumulative dose reaches a certain threshold and the subsequent exponential development of effects. In the case of lead, the blood concentration in most individuals may vary between 10 - 40 μg/100 ml, but clinical manifestations of lead poisoning occur when it reaches 70 μg/100 ml. If the exposure continues, so the blood concentration continues to climb, the clinical consequences quickly multiply.

11.1.2 ENVIRONMENTAL FACTORS

Numerous studies have shown that the metabolism of drugs by enzymes in the hepatic endoplasmic reticulum is enhanced by the pretreatment of animals with chlorinated hydrocarbons [9, 35, 59, 72, 217, 242]. Dicophane (DDT) lowers serum bilirubin in certain forms of congenital unconjugated

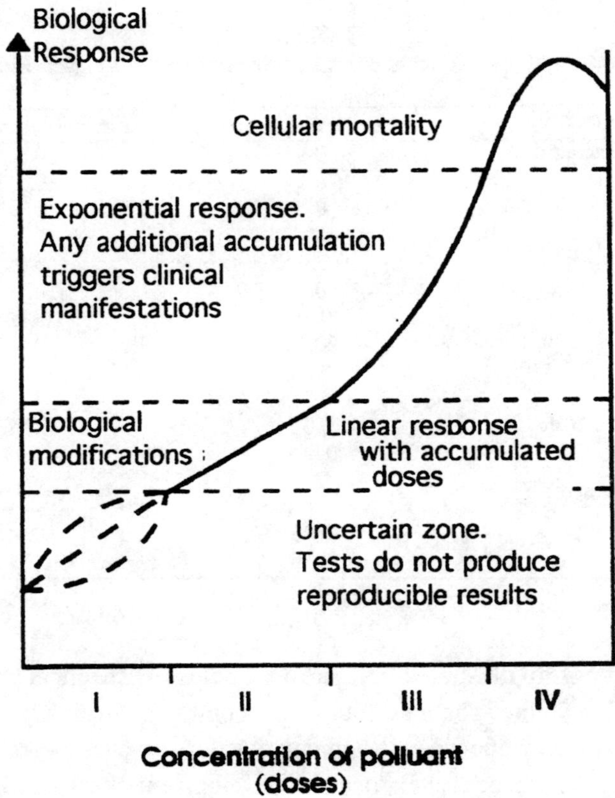

Figure 11.1. The relation between dose and biological response.

hyperbilirubinemia in which glucuronyl transferase is reduced. Also, blood levels of DDT were lower in patients being treated with the anticonvulsive drugs phenobarbitone or diphenylhydantoin, indicating more rapid metabolizm of DDT.

The interactions between the induction of biotransformation enzymes by drugs and chemical pollutants are not simple to understand. All xenobiotic compounds, including chemical pollutants, participate in the biotransformation process with varied rates depending on their structure and, especially, their dose. For example, polyaromatic hydrocarbons (PAH) induce enzymes in nonhepatic tissue more rapidly than DDT or phenobarbitone. As

Bick and Fishbein [25] observed, the absence of additive effects on enzyme induction suggests that both xenobiotics are metabolized by the same mechanism. An additive effect following a combined maximal stimulatory dose is seen only when the mechanism of induction for each compound is different, as with phenobarbital and certain PAHs. The changes observed after administering phenacetin and DDT, namely inhibition of induced enzyme activities with no marked increase in cytochrome P_{450} content, were considered a toxic response. This pattern is similar to that seen after intraperitoneal administration of well established hepatotoxic agents like carbon tetrachloride, thioacetamide, or chloroform.

Many substances in the environment, such as PAH, certain chlorinated insecticides, and drugs, stimulate MEOS and as a consequence affect drug therapy. It seems that an additive inductive effect can be expected only when the mechanism of induction is different, as with drugs or DDT or PAH. Concomitant exposure to metals and PAH or other organic chemical pollutants may produce additive effects.

By a close examination of the metabolic pathways of drugs, accidents due to cross induction may be avoided. Interactions can prolong or reduce the duration of action of the drug. Knowledge of these mechanisms and the individual's tolerance can reduce the possibility of a toxic reaction or lead to altering the treatment scheme.

Ionizing radiation or *radioactive fallout* is another source of pollution. Fortunately, this source of contamination is scarce, except in nuclear accidents like Chernobyl. There are several publications that demonstrate the interference of chemical and radioactive pollution [62, 212, 248, 269]. Both decrease the resistance of the body. They act by different initial mechanisms, but end up altering the same biochemical defense mechanism; glutathione directly or indirectly by decreasing NADPH biosynthesis. As discussed earlier, glutathione is a very effective detoxifying system. In addition, glutathione and the associated enzymes are the main natural defensive system against radiation.

Table 11.2

Metabolic processes influenced by human genetic polymorphism

1. Hydroxylation of xenobiotics by biotransformation enzymes
2. Spontaneous and chemically-induced carcinogenesis (aryl hydrocarbon hydroxylase)
3. Teratogenesis and senescence
4. Metabolism of drugs and ethyl alcohol
5. Biosynthesis of NADPH and of cholesterol

11.1.3 ORGANISM-DEPENDENT FACTORS

These are the most studied factors because of their great influence on the actions of drugs and toxic compounds [151, 163, 205]. Genetics, sex, age, body weight, circadian rhythm, stress, etc. are all organism-dependent factors.

Individual tolerance to exposure of chemical or infectious agents at dosages near the toxic threshold shows a wide range of variation. This variation is mostly the result of genetic polymorphism. As seen in table 11.2, most metabolic processes involved in the detoxification of xenobiotics are greatly influenced by genetic factors. Chemical sensitivity is also influenced by genetic polymorphism, but no extensive studies have been done [25, 121, 260].

Age strongly influences both sensitivity and resistance to chemical pollutants and, consequently, adaptation. *Newborns* are the group of the population most vulnerable to chemical pollutants due to their possession of an underdeveloped antioxidant system. In addition, many chemical pollutants can cross the placenta, increasing the degree of exposure. Aged people have a sensitivity to chemical pollutants similar to children because of decreased metabolic rate and decreased levels of antioxidants in the blood and tissues [39, 65, 299].

It has been claimed a number of times that either men or women have a higher resistance to chemical pollutants. A possible role of steroid hormones has been suggested, but no clear-cut studies have been done.

Physical effort produces several biologic actions including increased heart rate, faster breathing, metabolic acidosis, and an increase in blood amino acids and triglycerides. Fatigue increases the blood level of catecholamines and as a result decreases the body's resistance, making it more susceptible to the effects of chemical pollutants.

Stress is one of the most important factors as it decreases the resistance of the organism. Regardless of the type of stress (physical, emotional), the organism reacts with several biological and biochemical mechanisms: release of catecholamines and glucocorticoids, modification of carbohydrates and lipids, and altered blood circulation. The subsequent mobilization of glucose and triglycerides provides for the increased energy requirement of a stressed organism. In this condition, the organism is more vulnerable to the penetration of chemical pollutants. As was presented in table 1.1, there is biochemical evidence for emotional and other stress favoring lipid peroxidation in the heart.

Continuous stress favors an increased blood flow in the coronary vessels (catecholamine release) that, after a lengthy period of time, induces a decreased resistance of the vessel walls and favorable conditions for athersclerosis. The alternating contraction and relaxation of smooth muscle within the structure of the vessel wall eventually produces structural modifications, including the loss of creatine phosphokinase from the smooth muscles and its release into the blood. Another biochemical consequence is the release of lipids, which can be peroxidized and included in the formation of atherosclerotic plaques. The release of lipid is a nonspecific response of the body to stress, physical effort, and emotion. The principle mechanisms and biochemical consequences of stress are presented in figure 11.2.

Earlier we discussed the metabolism of drugs and chemical pollutants in the general context of xenobiotic compounds. There are several drugs (all tranquilizers) that are strong enzyme inducers that strongly interact with other chemicals, either stimulating or inhibiting metabolism depending on the chemical structure. The problem of ethyl alcohol has been presented. In addition to alcohol, environmental PAH and metals (Cd) can also strongly induce biotransformation enzymes, influencing the metabolism of drugs. In the same way, smoking contributes to an increased induction of biotransformation enzymes, influencing the metabolism of chemical pollutants and drugs [205, 252, 259].

Pregnancy can result in a decrease in glucuronic acid conjugation and in biotransformation of xenobiotic compounds in general because of the inhibitory effects of increased female hormones. These hormones act at the level of glucuronyl transferase and glutathione transferase [37].

Consequently, the second phase of biotransformation (glucuronic and sulfonyl conjugation) are affected. These modifications make the pregnant woman

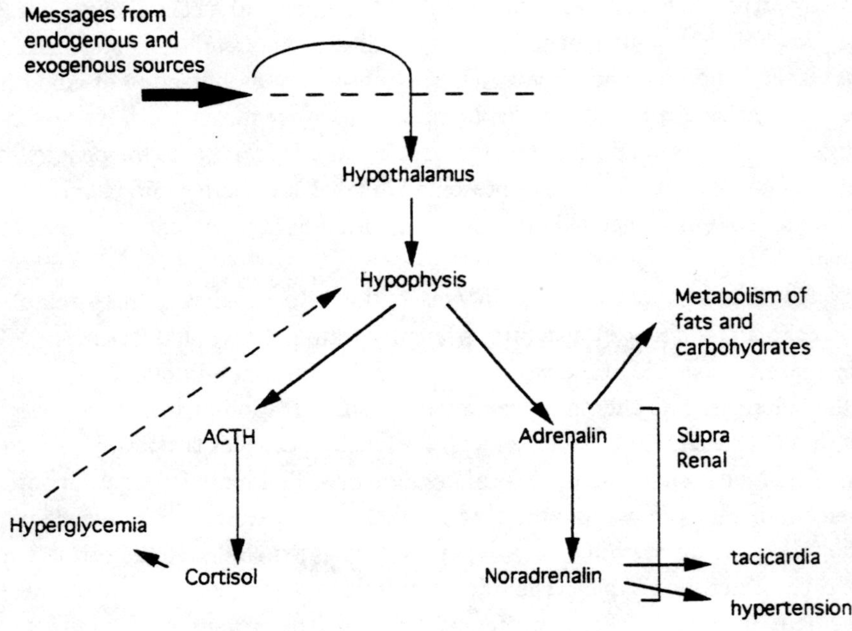

Figure 11.2. The body's main reactions in response to stress.

more sensitive to the action of chemical pollutants and xenobiotics.

11.1.4 NUTRITION

The role of nutrition and *foods* has been mentioned in other places in this book. Nutrition may be regarded as having a double effect on this subject. Food may be a source of contamination or a way to treat or prevent the consequences of pollution. In this section, we deal with both aspects and what might be considered a third: the introduction of chemical pollutants by cooking.

a. *Food as a source of protein.* Amino acids and proteins influence the biosynthesis of biotransformation enzymes and glutathione and consequently biotransformation and detoxification. The influence of a source of protein (casein) associated with a typical enzyme inducer (phenobarbital) is presented in table 11.3. The increase in amount of biotransformation enzymes along with protein is clear. One of the consequences of this is the modification of the lethal dose of a xenobiotic.

Table 11.3

The influence of diet and phenobarbital on some components of biotransformation enzymes. After Birnbaum [25].

Biotransformation Component	Casein			
	5%	5% + PB	20%	20% + PB
Liver wt (g)	7.0±0.4	7.9±0.1	5.2±0.4	6.5±0.2
Microsome protein (mg)	6.2±0.5	10.6±0.6	13.0±1.5	15.2±1.3
Cytochrome P_{450} (nmoles)	0.15	0.54	0.48	1.24
Km constant (mM)	0.12	0.19	0.14	0.32

PB = phenobarbital

Table 11.4

The effect of diet on rats exposed to insecticides. Numbers are lethal dose (LD_{50}) expressed in mg/kg. After Greenle and Polland [98].

Insecticide	Casein	
	4%	30%
Parathion	5	37
Malathion	766	1407
Carbetox	89	575

As shown in table 11.4, an increased intake of protein may increase the resistance of rats to insecticides by 2 - 5 fold. Alfthan [3] and Gue [227] investigated the effect of feeding charcoal-broiled beef on the *in vivo* metabolism of drugs in man. They found that feeding the charcoal-broiled beef diet for 4 days markedly lowered (4 fold) the plasma concentration of phenacetin due to an increased rate of metabolism. A high carbohydrate diet depresses the rate of metabolism of chemical pollutants.

b. *Food as a source of lipids*. Lipids are a vital and much discussed component of food. Cholesterol and polyunsaturated fatty acids (PUFA) are

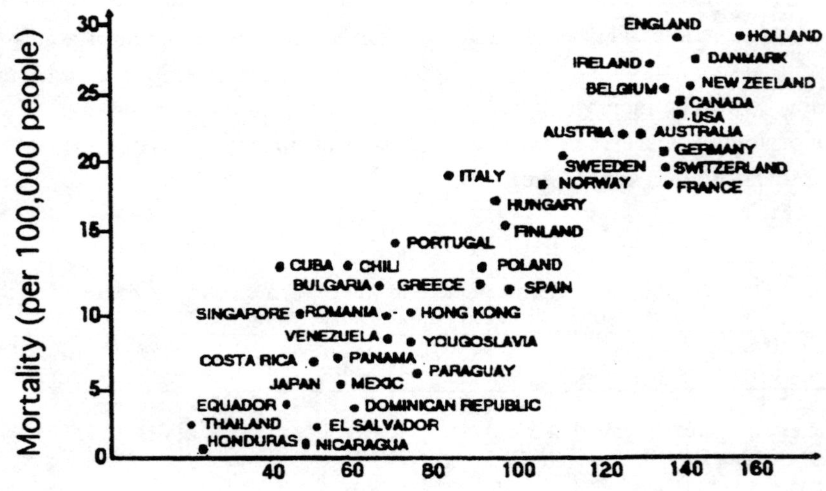

Figure 11.3. Relationship between mortality and fat-rich diet in various countries. FAO report, 1985.

the most debated groups of food components relating to obesity, atherosclerosis, and other diseases [7, 56, 109, 185, 237]. PUFA are only available to the organism by diet, making them similar in this respect to vitamins. PUFA are also very vulnerable to attack by oxygen free radicals, forming lipid peroxides and lipid free radicals.

As seen in figure 11.3, there is a definite, but still debated relationship between the frequency of breast and colon cancer and the amount of fat in the diet of people in countries around the world. The link between a high fat diet and the metabolism of chemical pollutants may be in the enhanced biosynthesis of steroid hormones, biliary acids, or prostaglandins that then influence the biotransformation of xenobiotics. The final word has not yet been said regarding the role of lipids in health.

The data presented in table 11.5 are quite amazing as they show that a diet rich in ocean fish and other marine life can protect against cardiovascular disease. The Eskimo and Japanese diet is rich in n-3 PUFAs (containing δ-linolenic acid). The n-3 PUFAs may have a form of antioxidant activity by virtue of their sensitivity to free radicals. They react with free radicals more easily than other important molecules (such as DNA), possibly removing them before they can react with DNA. N-3 PUFAs also enhance the

Table 11.5
Frequency of mortality from cardiovascular disease (CVD) and the PUFA content of the diet. After P. Weber [290].

Country	PUFA content (% arachidonic acid in platelet membrane)	PUFA quality (n-6/n-3)	Mortality by CVD (% of total)
Europe, USA	20 - 26	50	40
Japan	18 - 22	12	12
Greenland Eskimos	8 - 9	1.2	7

biosynthesis of high density lipoproteins (HDL).

c. *Cooked foods* as a source of chemical pollutants. The glycation hypothesis (glucose as a mediator in aging) was proposed by Carami in 1985 based on the observation of diabetes-related pathogenesis. Among the major reactions involved in some styles of cooking are certain nonenzymatic reactions, such as Maillard's reaction, involving the interaction of reducing sugars with pyrolyzed amino acids. In 1890, L. C. Maillard showed that heating glycine with glucose results in the formation of a brown-colored complex. As later demonstrated by B. S. Kristal and B. P Yu [149] of the University of San Antonio, the Maillard reaction is actually a series of complex reactions that involves initiation, propagation, and termination steps similar to those seen in free radical and lipid peroxidation reactions. All sugars (glucose, fructose) containing a aldehyde or keto group can react at 110° C with the epsilon amino group of amino acids. This results in a large range of colored products following condensation and polymerization reactions (Amadori products, Schiff bases), especially if iron or copper metals are present. Thus, the formation of Maillard products may not only generate free radicals, but their accumulation is enhanced by free radicals. Elevated glucose levels in man causes peroxidation of membrane lipids and enhanced membrane osmotic fragility. Increased glycation of hemoglobin, plasma proteins, and the lens of the eye also occur when the glucose level in the blood is increased, which is the case in severe diabetes mellitus.

The consequences of the Maillard reaction for cooked food are also very complex. As evident from the recipes in cook books down through the ages, Maillard browning reactions serve to enhance the sensory qualities of many dishes; the aroma of warm bread crust, roast meat, or freshly ground roast coffee are typical examples. The food industry is greatly interested in controlling the production of these characteristic aromas obtained in cooking, baking, roasting, and grilling. When bread or meat is overheated (above 120° C) on its surface during baking or roasting, respectively, a bitter flavor develops. Mixtures of sugars and amino acids also form bitter flavors when heated under these conditions. The most strongly bitter tastes in these experiments are obtained with proline, an abundant amino acid in collagen rich meat, and acrolein.

Since the introduction of the Ames test for mutagenicity, a wide range of foodstuffs, aroma concentrates, and other food additives have been tested [47, 62, 260]. Creatinine and tryptophan (present in meat) react with sugars at high temperatures, forming carcinogenic products. If the meat is heated above 130° C, the concentrations of these products is in the low ppb range. When the cooking temperature is limited to boiling, the concentration of these products is below the detection limit.

Secondary amines present in food stuffs can react with added nitrite during manufacture. But the Ames tests did not provide sufficient data on the carcinogenicity of these products, except for those previously discussed. For many years, sulfite was used as an additive to inhibit the formation of Maillard products in food. The entire process of the Maillard reaction is so complex and has so many confusing pathways leading to so many products (some even possessing antioxidant properties) that much effort will be needed to evaluate all the products. The presence of Maillard reactions in the human body has been recognized, but the consequences are as yet unknown.

Another source of chemical pollutants is the *synthetic steroid hormones*. These are used in several countries as growth hormones in animals. They are also used in contraceptives and for hormone replacement therapy to reduce osteoporosis and other post-menopausal problems. Diethylstilbestrol a sterol commonly used in growth hormone. Once it enters the body, diethylstilbestrol is hydroxylated in various positions. The resulting products may inhibit the biosynthesis of neurotransmitters, such as dopamine and noradrenaline, by acting at the level of the limiting enzyme, tyrosine hydroxylase. Oxygen free radicals are also formed. These synthetic hormones alter the biosynthesis of prolactin and natural steroid hormones. The formation of cholesterol epoxide is also favored. This compound possess mutagenic properties and has been

Table 11.6
Effect of Brussels sprouts and cabbage on intestinal metabolism. After
Chonney [42].

Diet	Metabolism of:		
	Phenacetin	7-Ethoxycoumarin	Benzopyrene
Control	1.65±0.28	68±5	13±2
Cabbage	6.07±0.75	1,073±177	428±109
Brussels sprouts	10.88±0.37	2,904±375	1,112±283

found in breast milk in significant concentrations (800 nmolar) [92, 195, 229].
Dozens of synthetic chemicals have been found to act like hormones. They
can affect the brain, immune system, or reproductive system. Bisphenol-A
(found in polycarbonate plastic), nonylphenols (found in PVC plastic and as a
breakdown product of industrial detergents and pesticides), and dioxin (a
common combustion product) can mimic the function of estrogen. PCBs
(used in electrical transformers) can act like thyroid hormones. Vinclozolin
(used as a fungicide) can bind to cellular testosterone receptors. The
involvement of estrogen like compounds in the reduction in male fertility is
currently under very active debate.

   d. *Food as a source of antioxidant vitamins*. There is an abundance of
data from experiments that show an antioxidant rich diet significantly
influences the metabolism of xenobiotic compounds. For example, Chonney
[42] demonstrated the effects of feeding rats cabbage, a vegetable rich in
vitamin C and minerals (table 11.6). The rats were fed a 25% dried vegetable
powder in a nutritionally complete synthetic diet for 7 days.

   Several studies [2, 185, 227] performed on 25,000 volunteers from
Washington State showed:

- High serum β-carotene levels were strongly associated with protection
  from lung cancer, melanoma, and bladder cancer and had a lessor effect
  against rectal cancer. However, more recent reports contradict this
  finding.
- High serum vitamin E levels were associated with protection from lung
  cancer;
- Low levels of serum lycopene (an antioxidative carotene) were strongly
  associated with skin cancer.

The localization of chemical or occupationally induced cancers is related to the route of penetration of the chemical: lung, skin, gastrointestinal tract. Recently, it was shown that the selenium content of the diet and blood had no conclusive relationship towards cancer of the gastrointestinal tract, breast, or lung [165, 224, 285].

The protective effect of antioxidant vitamins is one of the major areas of current research and is the subject of national and international programs of epidemiological studies. The purpose of this intense research is related to the role of antioxidants in preventing the so called degenerative diseases (cardiovascular disease, cancer, pancreatitis, and diabetes) that occur with high frequency in developed, industrialized countries.

Another subject is related to the *role of diet* in preventing gastrointestinal tract problems, including colon cancer. The fiber found in cereals, fruits, and vegetables are now recognized as necessary for maintaining proper intestinal peristalsis. Fiber contains cellulose, nondigestable oligosaccharides, pectins, pentosanes, and other partially known compounds. Fiber increases the rate of peristalsis, decreasing the time food (and chemical pollutants) remain in the intestine. By decreasing the contact time with the intestine the absorption of harmful substances is reduced and there is less time for various compounds to become involved in bacterial fermentations due to intestinal flora. Through this action fiber may be effective in preventing the development of colon and rectal cancer.

Recently, at the ENEA symposium, D. Piero from the University of Florence showed that populations consuming high levels of starch and fiber have a lower incidence of colon cancer and that animals fed a diet containing 5% olive oil had a lower aryl hydrocarbon hydroxylase activity that might be correlated with reduced genotoxic activity (mutation and cancer). Cellular proliferation is known to be an important regulatory control on the expression of chemically induced gastrointestinal tumors and is itself dependent on the variations of dietary components [185, 255, 297, 300].

The role of diet in the modification of the resistance of people to disease is far from completely understood. The continuing growth of the literature concerning ways to prevent the consequences of exposure to dangerous pollutants and the development of degenerative diseases is proof of the intense interest in this area.

11.2 TERATOGENIC EFFECTS OF CHEMICAL POLLUTION

Many chemicals are overtly or potentially harmful, causing subacute effects on reproduction. Reproductive effects are rarely the primary or only toxic manifestation of a chemical, but occasionally an embryo or fetus is the main or only victim of an environmental chemical, indicating the embryo may have extraordinary sensitivity to teratogens. The embryo or fetus within the mother's body often is exposed to a much reduce fraction of the mother's exposure, but such small amounts may be sufficient to cause damage.

Homeostatic mechanisms are able to mitigate the dosage that would otherwise reach the embryo. The maternal liver and kidneys largely protect against the penetration of chemical pollutants or their metabolites. Some chemical pollutants may be transferred from the plasma into storage depots, thus tending to minimize the peak concentrations in plasma. For example, DDT and other lipid-soluble chemicals enter the fat stores and bones. The placenta is often referred to as a barrier, but actually most unbound chemicals pass to the embryo by simple diffusion. The fate of chemical pollutants within the embryo is poorly known, but protein binding has been demonstrated. Incorporation into embryonic tissues is known to occur with certain antimetabolites that are capable of replacing physiological precursors for biosynthesis. Presently, the known causes of developmental defects in man are:

- Genetic transmission; 20%
- Chromosomal aberrations; 3 - 5%
- Ionizing radiation (including radioactive fallout); < 1%
- Infections (varicella, cytomegalovirus); 2 - 3%
- Maternal metabolic imbalances; 1 - 2%
- Drugs and environmental chemicals; 4 - 6%
- Unknown; 65 - 70%.

This data indicates that all currently known environmental factors combined do not account for more than 10% of birth defects [122, 147, 288]. Of course, some environmental accidents (Minamata, Sevasso, Chernobyl) modify this data, but these events are restricted to specific areas. It should be emphasized that the effect of chemical pollutants are thought to derive from combinations and interactions among two or more factors, all of which may not yet be identified. Although the effect has not yet been demonstrated in man, multiple environmental factors at low doses may interact to cause

Table 11.7
Drugs and chemicals known to be toxic to human embryos.

Compound	Dose (μg)	Type of defects
Thalidomide	Thousands	Reduced limb development
Androgenic hormones	Hundreds	Virilization of females
Folate antagonists	Tens	Spontaneous abortions
Methyl mercury	Dozens	Neuro-sensory disorders
Maternal alcoholism		Growth retardation, impared development of nervous system and eyes
Alkylating agents		Intrauterine death
Neurotropic and anticonvulsant drugs		Heart and other organs

additive or potentiating effects on development.

Table 11.7 lists some drugs and chemicals known to cause injury to human embryos. To this list should be added another list containing compounds tested for embryo toxicity in animals. Surprisingly, only dioxin, dimethylsulfoxide, and some metals (mercury, lithium, lead, cadmium, and arsenic) are found to possess high embryo toxicity in animals. The chemicals studied make a long list, including fungicides, herbicides, insecticides.

Studies on embryo toxic compounds are far from complete and, especially for man, the conclusions may be very difficult to obtain.

11.3 BIOLOGIC DOSIMETRY AND BIOMONITORS

There are known marker enzymes for toxicity and methods for detecting DNA-damage products. These are components of biologic dosimetry, which is the estimation of the physically, chemically, or biologically meaningful dose received from exposure to chemical, bacterial, or radiological agents.

The use of the term "biological" can refer to either the biologic origin of the sample to be measured, or to the biological mechanism through which the dose elicits relevant effects. When used in a reasonably limited setting for a practical purpose, biological dosimetry should relate the dose to a biological response observed in the subject.

To investigate a chemical pollutant that is absorbed, transported and metabolized, biological dosimetry begins with a measurement of the chemical and its distribution in various parts of the body (partitioning). This part of the study may be designed as chemical or pharmacological dosimetry. At the end of the chemical's biotransformation, the estimation of protein- or DNA-adducts provides an indication of the molecular consequences of exposure to the xenobiotic. The *in vitro* counting of lost cells or those with mutations is included in the field of biological response, as would be later observations of preneoplastic changes. The final components of biological dosimetry involve epidemiological studies, altered life expectancy, or species survival. Such studies allow us to know that, for example, patients with mesothelioma, lung pneumoconiosis, angiosarcoma of the liver, or juvenile thyroid cancer have been exposed to asbestos, vinyl chloride, or ionizing radiation.

In 1988, the Commission of European Communities launched a research project titled Biomonitoring of Human Populations Exposed to Genotoxic Environmental Chemicals. This project was intended to assist the development of population monitoring systems designed to quantify the exposure to potential mutagenic chemical in the environment and to detect possible early effects. The actual approach is based on various aspects of methodological development for:

a) The identification and measurement of DNA adducts by P^{32}-labeling methods, immunologic assays, and advanced mass spectrometry techniques;

b) The detection of hemoglobin and plasma protein adducts by gas chromatography-mass spectrometry;

c) The detection of chromosomal aberrations, sister chromatid exchange, and micronuclei in lymphocytes;

d) Quantitative determination of the hypoxanthine guanine phosphoribosyl transferase mutation frequency in lymphocytes and mutations in the hemoglobin gene (glycophorin A genes).

Such techniques and others, while possible, are expensive and it is difficult to extrapolate the results [20, 80, 119, 240].

For practical, rapid, and inexpensive measurements of chemical pollution, biomonitors are widely used. The most sensitive biological instruments for measuring chemical pollutant intake are bryophytes or mosses.

Bryophytes are plants lacking true roots, acquiring water mainly through their surfaces. This kind of water uptake is favored by the high surface to volume ratio of these plants. This makes the water, and hence chemical pollutants or metals, intake in mosses mainly independent of the substrate composition and concentration. Mosses exhibit ion exchange properties. For example, *Hylocomium splendens* favors the uptake of metals in the following sequence: Pb^{+2}, Cu^{+2} > Co^{+2} > Zn^{+2}, Mn^{+2}. Even in the presence of high concentration of competing ions such as Mg^{+2}, Ca^{+2}, K^+, Na^+, copper and lead ions were completely absorbed. This leads to the conclusion that they are easily absorbed from rain water. The concentrations in *Hylocomium* in a contaminated area may be as high as 0.2 mg/g for lead, 0.1 mg/g for copper, 2 µg/g for cadmium, and 0.4 mg/g for zinc. Even higher concentrations were found in *Brymum argentum* in urban areas. The maximum content (per gram dry weight) were: 5 mg for lead, 2 mg for copper, 60 µg for cadmium, and 5.8 mg for zinc.

The wide variations in morphology, growth, and cation exchange capacity shown by different species of mosses is largely responsible for the equally large variations in concentrations of heavy metals found in different species.

Species of lichens are known to be sensitive to various air pollutants. Field and laboratory studies have demonstrated that lichens are extremely sensitive to sulfur dioxide, heavy metals, or hydrogen fluoride. The moisture content of lichen is extremely important in determining its response to SO_2.

Cladonia alpestris exhibits a 45% decrease in photosynthesis after a 6 hour exposure to 0.75 ppm SO_2. *C. alpestris* is sensitive to concentrations of:

- 0.56 mg SO_2/m^3 and perhaps as low as 0.25 SO_2/m^3
- 30 - 40 µg floride/g dry weight
- 450 µg Cd/g

As shown by T. Nash of the University of Tempe, Arizona, lichens may accumulate astonishingly great amounts of heavy metals: 90,000 µg Fe/g lichen, 5,000 µg Cd/g lichen, and 17,000 µg Pb/g lichen. In none of these studies has toxicity been demonstrated. In all of these cases, SO_2 was also present in high concentrations and may well be the most important toxic factor.

In contrast to mosses and lichens, fungi can reflect the heavy metal content of the soil. The mycelia deeply penetrate the soil throughout the year, while the fruiting body is only exposed to the air for a short time. The fruiting bodies of *Agaricus* are highly susceptible to the accumulation of mercury and cadmium.

Certain birds, such as blackbirds, magpies, goshawks, and canaries (for Co), are among the few vertebrate species that can be used as biomonitors.

For monitoring soil status earthworms were found to exhibit fairly high concentrations of heavy metals and arthropods are suitable for biomonitoring soil metal content in forest ecosystems.

In aquatic marine ecosystems, mollusks (*Mytilus edulis*) and macroalgae fulfill many indicator requirements, especially *Fucus* in temperate water and *Sargassum* in subtropical and tropical waters.

Of these examples, mosses and lichens have the most widespread use in biomonitoring chemical pollution.

A different type of example of this is algal blooms that are toxic for animals and humans. The increase in algal blooms is a consequence of an increased nutrient load, particularly nitrates and phosphates. Blooming algae contain high amounts of neuro- and hepato-toxins as well as contact irritants.

11.4 BIODEGRADATION OF POLLUTING CHEMICAL SUBSTANCES

A wide variety of chemical pollutants present in the soil, fresh water, and oceans are toxic or can be converted into toxic compounds. In recent years, special interest has been given to the biodegradation of compounds present in nature through the activity of microbes present in the environment. The microbial population primarily acts through enzymatic activity, but nonenzymatic reactions may be involved as well. Microbes convert organic substances to other organic compounds using the released carbon to construct new cell material. The process takes place with the release of energy and the bacterial population increases in number. Mineralization is a typical process of microbial growth [41, 240].

To date, evidence for biodegradation has been obtained for insecticides, herbicides, and detergents. But these studies are hindered by the lack of knowledge of these processes in nature. Quite often the introduction of new chemical compounds into a culture medium does not induce a corresponding and proportional growth in bacterial population, even when there is evidence the new compound is being metabolized. Such co-metabolism of some chemical compounds is frequent. This is the case for synthetic compounds for which the microbial population lacks the specific enzyme necessary to metabolism. The evolution of new enzymes is a time consuming process in which all members of a given population may not be involved.

The enzymes involved in the biotransformation of organic compounds have quite a low substrate specificity so that many times one substance can be metabolized through several enzymatic pathways. The same phenomenon takes place in a bacterial population with the added advantage of a faster growth rate allowing more plasticity and adaptability.

Though the biodegradation of organic substances is essentially a decontamination process, the inverse phenomenon is also possible; i.e., activation and increased toxicity, as is the case with the methylation of mercury and arsenic. In this case the resulting compounds are more toxic than the substrates.

Table 11.8 presents a portion of the reactions found in nature. It may be noticed that for some compounds, as is the case for DDT, several metabolic possibilities exist. If, during the process of degradation, the metabolism of a certain organic compound reaches a metabolite for which there is no adequate enzyme activity, the intermediate accumulates in the environment, often causing the inhibition of bacterial growth. In biotechnological processes as well as in the application of herbicides, insecticides and fertilizers, use of too high a concentration; i.e., above the scientifically established dose, causes inhibition and poisoning of the microbial cultures in the biosphere by blocking the enzymatic reactions involved in the chemical's degradation.

Table 11.8

Types of reactions involved in the biodegradation of polluting compounds, mostly organo-phosphorous compounds.

Catagory	Reaction	Example
Dehalogenation	$RCH_2Cl \rightarrow RH_2OH$	Propachlor (S, M)
Decarboxylation	$Ar_2CHCOOH \rightarrow Ar_2CH_2$	DDT (M)
Hydroxylation	$RCH_2R' \rightarrow RCH(OH)R'$	Carbofuran (S, N, W)
Epoxide formation	$RCH{=}CHR' \rightarrow$ $CCHO(O)CHR'$	Heptachlor (S, M)
Oxydations	$(AlcO)_2P(S)R \rightarrow$ $(AlcO)_2P(O)R$	Parathion (S, M)
Reduction of double bond	$Ar_2C{=}CH_2 \rightarrow Ar_2CHCH_3$	DDT (N, M)
Nitrogen metabolism	$RNO_2 \rightarrow RNH_2$	Sumithion (S, N)
Splitting ester	$RC(O)OR' \rightarrow RC(O)OH$	Malathion (N, M)
Reduction of disulfides	$RSSR \rightarrow 2RSH$	Thiram (S, M)

Reactions demonstrated in waste water (W), microbial culture (M), soil (S), and natural waters (N).

APPENDIX

Appendix I
Adult daily requirements of some vitamins and minerals

Vitamins (mg)		Metallic ions (mg)	
A (retinol)	0.8 - 1	Phosphorous	800 - 1,200
B1 (thiamine)	1.1 - 1.5	Calcium	800 - 1,200
B2 (riboflavin)	1.3 - 1.7	Magnesium	280 - 350
B6 (pyridoxine)	1.6 - 2	Zinc	12 - 15
Biotin	0.03 - 0.1	Iron	10 - 15
Folic acid	0.18 - 0.2	Selenium	0.055 - 0.07
Niacin	15 - 19	Copper	1.5 - 3.0
B12	0.002	Iodine	0.15
C (ascorbic acid)	60	Manganese	2 - 5
E (tocopherol)	8 - 10		
K	0.06 - 0.08		

Appendix II
Food content of some vitamins and minerals. Amounts expressed per 100
grams of food.

Food	Prot. (g)	Vitamins A	B1	B2	C	E	Minerals Ca	P	Fe	Se
Bread	9.6	0.3	0.15	0.20	—	—	50	200	2.5	0.67
Cow's milk	3.3	0.20	0.05	0.20	1.0	1.5	120	93	0.24	0.02 - 0.1
Beef	20	0.03	0.06	0.25	1.0	6.1	200	300	.030	0.356
Pork	25	0.03	0.18	0.03	1.0	—	15	220	0.40	0.022
Chicken	20	0.15	0.07	0.08	5.0	1.0	11	208	0.60	0.12
Eggs	13	1.5	0.17	0.20	—	1.0	60	250	0.07	0.175
Cheese	20 - 30	0.06	0.08	0.05	—	—	>700	>200	0.03	0.06 - 0.2
Boiled rice	7.8	—	—	—	—	—	2	30	0.03	0.383
Mushroomes	4	—	0.02	0.64	4	—	65	387	304	0.643
Rye bread	9	—	0.5	0.25	—	15	14	98	0.7	0.122
Wheat germ	27	0.30	2.6	0.50	2	14	24	150	1.3	0.070
Red pepper	1.3	2.2	0.14	0.05	125	7	70	880	7.5	0.35
Carrots	1.2	5.4	0.06	0.05	4	1.5	20	56	0.08	0.07
Raw cabbage	1.4	—	0.09	0.06	45	2	14	26	0.40	0.04
Cucumber	0.7	0.21	0.05	0.01	6	—	41	52	0.62	0.022
Endive	1.6	1.2	0.10	0.10	4	—	45	30	0.43	0.025
Cauliflower	2.4	1.3	0.10	0.10	4	—	10	22	0.33	0.012
Green beans	2.5	0.3	0.12	0.24	10	—	104	38	0.123	0.05
Turnip	1.1	—	0.05	0.06	24	—	122	60	1.2	0.08
Garlic	4.4	—	0.06	0.08	4	—	46	52	1.1	0.02
Horse radish	3.2	—	0.01	0.02	54	—	56	47	0.52	—
Lemon juice	0.9	0.12	0.05	0.02	65	—	42	35	0.07	0.02
Plum	2.1	0.27	0.20	0.55	10	—	96	75	1.3	0.07
Strawberry	0.8	0.3	0.2	0.1	35	—	22	11	0.6	—

INDEX

89. K. Furukawa, N. Kimura. *Environ. Health Persp. Suppl.* 103: 21-23, 1995.

90. P. R. Garcia, D. Martinez, M. Repetto. *Vet. Hum. Toxicol.* 37: 306-309, 1995.

91. D. Gemsa, H. G. Loser, E. Berlin. *Molec. Immunol.* 19: 1287-1292, 1982.

92. A. Ghoshal, T. Reishmore. In: *Nongenotoxic Mechanisms in Carcenogenesis.* J. T. Sease ed. Butterworth, London, 155-159, 1987.

93. J. H. Gill, C. A. Molloy, K. J. Shoesmith. *Cell Death Differ.* 2: 211-217, 1995.

94. P. F. Giengeuch, R. Thieu, M. Persmark. *Pharmacogenetics.* 5: S103-S107, 1995.

95. E. Gillepsie. *Nature.* 277: 135-136, 1979.

96. M. R. Goel, M. Sbarra. *Bull. Environ. Toxicol.* 40: 225-229, 1988.

97. W. J. Goldman. *Lancet.* 345: 326-328, 1995.

98. W. F. Greenle, A. Pollard. *J. Biol. Chem.* 254: 1105-1108, 1979.

99. L. S. Green. *J. Am. Col. Nutr.* 14: 317-324, 1995.

100. M. Grootveld, J. R. Hault, D. R. Lake. *Biochem. J.* 277: 499-504, 1986.

101. M. Grootveld,, J. M. C. Gutteridge. *Biochem. J.* 273: 459-467, 1991.

102. P. Gupta. A. Kar, P. K. Maiti. *Fresenius Environ. Bull.* 4: 333-335, 1995.

103. J. A. Gustafsson, E. G. Hrycay. *Archic. Biochem. Phys.* 174: 440-445, 1977.

104. L. E. Guray, J. Sostby. Toxicol. *Appl. Pharmacol.* 133: 285-289, 1995.

105. J. M. C. Gutteridge, B. Halliwell. *Biochem. J.* 243: 709-714, 1987.

106. M. C. Gutertez, L. Bucio, V. Sauza. *Hum. Exp. Toxicol.* 14: 324-334, 1995.

107. B. Halliwell, J. R. Hault, D. R. Blake. *FASEB* J. 2: 2867-2873, 1988.

108. B. Halliwell, J. M. Gutteridge. In: *Methods in Enzymology,* Vol. 186, L. Packer, A. N. Glazer, eds. p. 1-85, 1994.

109. B. Halliwell, J. M. C. Gutteridge. *Free Radicals in Biology and Medicine,* 2nd edition. Claredon Press, Oxford, 1994.

110. L. Hardell, M. Eriksson, L. Athlin. *Oncol. Reports.* 2: 749-753, 1995.

111. D. Harman. *Age.* 17: 119-146, 1994.

112. Y. Hashimoto. *Pharmacogenetics.* 5: S80-S83, 1995.

113. L. Heap, R. J. Ward, C. Abiaka. *Biochem. Pharmacol.* 50: 263-270, 1995.

114. M. Hedenborb, M. Klockhaus. *J. Clin. Pathol.* 40: 1190-1194, 1987.

115. D. L. Henshaw, P. A. Keitch, P. R. James. *Lancet* 345: 324-325, 1995.

116. V. P. Heinonen. *New Engl. J. Med.* 330: 1029-1035, 1995.

117. K. Hemminki. *Toxicology.* 101: 41-53, 1995.

118. H. Hewig, R. Rucker. *Toxicol. Lett.* 38: 225-228, 1987.

119. K. Hewincki, P. Vainio. In: *Monitoring Human Exposure to Carcinogic and Mutagenic Agents.* IRAC Scientific Pub., Nr. 78, Lyon, 1986.

120. B. Hillman. *Chem. Eng. News.* 69: 26-28, 1991.

121. D. S. Hill, K C. Bhujan. *Exp. Eye Res.* 39: 391-394, 1984.

122. IARC, *Environmental Carcinogens.* IRAC Scientific Pub., Nr. 78, Lyon, 1986.

123. IARC, *Directory of On-Going Research in Cancer Epidemiology.* IRAC Scientific Pub., Nr. 78, Lyon, 1986.

124. M. Ishizaki. *Environ. Sci. Tokyo.* 3: 173-186, 1995.

125. M. Isobe. T. Stone, E. Takabata. *J. Toxicol. Sci.* 20: 161-164, 1995.

126. IUPAC, Commission Agrochemicals. *Pure Appl. Chem.* 67: 1487-1587, 1995.

127. R. Jam, K. D. Mishra. *Pollut. Res.* 13: 375-380, 1994.

128. J. P. James, G. B. Quistad. J. *Agric. Food Chem.* 43: 2530-2535, 1995.

129. Y. M. Jansen, A. Barchowsky, M. Treadwell. *Proc. Natl. Acad. Sci.* U.S. 92: 8458-8462, 1995.

130. A. M. Jarabek. *Inhalation Toxicol.* 7: 927-946, 1995.

131. T. W. Jones, S. Orrenius, H. Thorwald. *Archiv. Toxicol. Suppl.* 9: 260-263, 1986.

132. K. Jung, B. Smith. *Clin. Chem.* 39: 757-759, 1993.

133. K. R. Kaderlick, F. F. Kadlubar. *Pharmacogenetics.* 5: S108-S117, 1995.

134. R. Kahl, U. Wolf. *Toxicol. Appl. Pharmacol.* 47: 217-220, 1979.

135. T. Karniyama, H. Miyakawa, J. P. Li. *Res. Com. Mol. Pathol. Pharmacol.* 88: 177-186, 1995.

136. A. Kappas, A. P. Alvarez. *Sci. Am.* 232: 22-27, 1975.

137. H. Kasai, S. Nishikimi. *Environ. Health Perspect.* 67: 171-174, 1986.

138. K. S. Kasprzak. *Cancer Invest.* 13: 411-430, 1995.

139. H. Kawahara, Y. Matsuda, S. Takase. *Alcohol,* Suppl. 1: 113-118, 1994.

140. R. A. Keer, W. J. Jakoby. *Science.* 237: 1183-1185, 1987.

141. S. Killinger, Y. Li, V. J. Balcar. *Neuroreport.* 6: 1290-1292, 1995.

142. D. E. Kizer, J. A. Claus. *Biochem. Pharmcol.* 34: 1795-1798, 1985.

143. C. D. Klaasen. *Drug Metabolism Rev.* 5: 165-170, 1976.

144. J. Klimczak, J. Kolawski. *Toxicology.* 32: 267-271, 1984.

145. N. Koga. *Biol. Pharm. Bull.* 18: 205-210, 1995.

146. H. S. Koren, P. A. Bromberg. *Intl. Arch. Allergy Immunol.* 107: 236-238, 1995.

147. M. Kovacs. *Biological Indicators in Environmental Protection.* Harwood, Prentice Hall, Engelwood Cliffs, NY, Chapters 2 – 3, 1992.

148. N. I. Krinsky. In: *Antioxidant Vitamins and Carotenes in Disease Prevention.* T. F. Slater, ed. Am. J. Clin. Nutr. Suppl. 1. 53: 238S-246S, 1991.

149. B. S. Kristal, B. P. Yu. *J. Gerontol.* 47: 107-110, 1992.

150. C. A. Krone, S. M. Yeh, W. T. Idawa. *Environ. Health Perspect.* 67: 75 80, 1986.

151. H. Kulkarn, K. R. Murty. *Xenobiotica.* 25: 773-810, 1995.

152. J. Lahdetie. *J. Occupat. Environ. Med.* 37: 922-926, 1995.

153. A. L. Landay, C. Jessop, J. A. Levy. *Lancet.* 338: 707-710, 1991.

154. J. C. Larsen. *Toxicology.* 101: 11-27, 1995.

155. R. J. Larson. *Environ. Toxicol. Chem.* 14: 1423-1442, 1995.

156. H. S. Lau, S. L. Coffing, H. Lee, R. G. Harvey. *Chem. Res. Toxicol.* 8: 970-978, 1995.

157. J. Leff, W. Carter. *Lancet.* 341: 777-779, 1995.

158. T. Leenadevi, K. V. Vasale, A. Payan. *Mycotoxin Res.* 11: 2-8, 1995.

159. D. F. Lewes, C. Ioanides, D. V. Parke. *Biochem. Pharmacol.* 50: 619-625, 1995.

160. A. P. Li. *Toxicol. Appl. Pharmacol.* 57:55-60, 1981.

161. C. S. Lieber. *Sci. Am.* 234: 3-8, 1977.

162. Y. Lin, D. Yamieson. *Pathophysiology.* 2: 9-16, 1995.

163. M. L. Lindbohm. *J. Occupat. Environ. Med.* 37: 908-912, 1995.

164. L. Liu, P. G. Wells. *Toxicol. Appl. Pharmacol.* 134: 424-428, 1995.

165. O. R. Livotto, M. Duren. *Free Rad. Biol. Med.* 19: 431-440, 1995.

166. E. L. Loechler. *Mol. Carcinog.* 13: 213-219, 1995.

167. L. Z. Lorscheider, M. J. Vimy, A. O. Summers. *FASEB J.* 9: 504-508, 1995.

168. K. Lowrey, A. Glende. *Biochem. Pharmacol.* 30: 135-138, 1981.

169. C. Lu, H. Yagi, D. M. Jerina. In: *Applied Molecular Biology in Environmental Chemistry.* R. A. Mineaz, R. A. Lewes, eds. CRC Press. Boca Raton, FL. 35-44, 1995.

170. B. Lund, B. Lund. *J. Biol. Chem.* 270: 20895-20897, 1995.

171. T. Lundh, B. Akesson. *J. Occupat. Environ. Med.* 52: 478-483, 1995.

172. J. Lunec, S. P. Alloran. *J. Clin. Invest.* 76: 2084-2088, 1984.

173. M. C. Macleod. *Carcinogenesis.* 16: 2009-2014, 1995.

174. K. R. Mahaffey. *Inhalation Toxicol.* 7: 1019-1025, 1995.

175. V. M. Maher, R. H. Chen, J. McCormick. *Appl. Molec. Biol. Environ. Chem.* 59-67, 1995.

176. P. F. Mannanioni, M. G. Bello, S. Raspanti. *Adv. Prostagl. Thromb. Leuk.* Res. 23: 215-217, 1995.

177. S. L. Markund. In: *Oxygen Radical in Chemistry and Biology.* W. Bors, M. Saroyan, eds. W. de Gruif, NY, 765, 1984

178. J. L. Marx. *Science.* 235: 528-531, 1987.

179. M. A. Mayorga, J. Januskiewicz, B. E. Behnert. *Polym Mater. Sci. Eng.* 71: 181-183, 1994.

180. J. McCord. *Proc. Soc. Exp. Biol. Med.* 209: 112-117, 1995.

181. A. Meister. *J. Biol. Chem.* 263: 17205-17208, 1988.

182. A. Meister. *J. Biol. Chem.* 269: 9397-9400, 1994.

183. F. Z. Meerson. *Adaptation, Stress and Prophylaxis.* Springer. Berlin, 1985.

184. P. Merry, M. Grootveld, J. Lunec. *Am. J. Clin. Nutr. Soc. Suppl.* 53: 362S-369S, 1991.

185. F. L. Meyskens. In: *Vitamins, Nutrition and Cancer.* K. N. Prasad, ed. S. Krager, London, 1984, p. 266.

186. A. Miller, E. C. Miller. *J. Natl. Cancer Inst.* 47: 5-10, 1971.
187. M. J. Miller, D. C. Parmlee, F. F. Kadlubar. *Chem. Biol. Interact.* 93: 221-234, 1994.
188. E. G. Minnaugh. *J. Phar. Exp. Theoret.* 226: 806 809, 1984.
189. N. Mistry, K. Herbert, M. D. Evans. *Biochem. Soc. Trans.* 23: 482-485, 1995.
190. S. Moncada, R. M. Palmer. *Proc. Natl. Acad. Sci. U. S.* 33: 9164-9167, 1986.
191. N. A. Molfino, S. C. Wright, I. Katz. *Lancet.* 338: 199-203, 1991.
192. J. Molnar, I. Beladik, K. Domonkos. *Neoplasma.* 28: 1-5, 1981.
193. R. A. Morgan, G. T. Fenwick. *Lancet.* 336: 1492-1494, 1995.
194. J. L. Muderly. *Toxicol. Environ. Chem.* 49: 167-168, 1995.
195. M. F. A. Muller, V. Markopoulos, H. M. Bolt. *Exotoxicol. News.* 2: 68-73, 1995.
196. N. Murata, H. Hiroyuki. *Carcinogenesis.* 16: 2251-2253, 1995.
197. J. F. Naarala, J. J. Loikkanen, K. M. Savolainen. *Free Rad. Biol. Med.* 19: 689-693, 1995.
198. M. Nagao, Y. Fujita, T. Kosuge. *Environ. Health Perspect.* 67: 89-93, 1986.
199. C. P. Nair, A. K. De, R. Darad. *J. Exp. Biol.* 33: 275-277, 1995.
200. M. Nakatani, R. Inatani, H. Ohta. *Environ. Health Perspect.* 67: 135-138, 1986.
201. A. A. Nanji, B. Griniuvene, K. Yocub. *Proc. Soc. Exp. Biol. Med.* 210: 12-19, 1995.
202. National Research Council: *Multiple Chemical Sensitivity. Addendum. Biological Markers in Immunotoxicology.* Washington DC. Natl. Acad. Press. 1992.
203. B. Nehru, A. Yier. *J. Environ. Pathol. Toxicol. Oncol.* 13: 265-268, 1995.
204. J. Neve. *Med. Hygiene.* 51: 741-745, 1993.
205. A. Newman. *Lancet.* 345: 296-297, 1995.
206. E. Nieboer, B. L. Gibson. *Environ. Res.* 3: 29-81, 1995.
207. P. Nolan, P. Masel. *Lancet.* 345: 67-68, 1995.
208. L. W. Oberley. *Archiv. Biochem. Biophys.* 254: 69-73, 1987.
209. F. Oesch, J. G. Hengstler. *Pharmcogenetics.* 5: S118-S122, 1995.
210. S. Okada, S. Yamazake. *Biochem. Biophys. Acta.* 922: 28-34, 1987.
211. R. Olinescu. *Peroxidation in Chemistry, Biology, and Medicine.* Edit. Stiintifica. Bucharest. 1982.
212. R. Olinescu, R. Alexandrescu. *Rev. Roum. Biochim.* 22: 193-298, 1985.
213. R. Olinescu, S. Nita. *Rev. Roum. Med. Int.* 30: 201-208, 1992.
214. R. Olinescu, S. Nita. *Rev. Roum. Med. Int.* 31: 109-112, 1993.
215. N. Orenterich, N. Krieger. *J. Natl. Can. Inst.* 86: 589-599, 1994.
216. U. Ozkutlu, J. P. Long, J. Q. Cannon. *Arch. Int. Pharmcodyn. Therap.* 329: 331-342, 1995.
217. S. Padella. *Inhalation Toxicol.* 7: 903-907, 1995.
218. D. N. Pathak, G. Levay, W. J. Badell. *Carcinogenesis.* 16: 1803-1808, 1995.
219. H. E. Paulsen, S. Loft. *Toxicology.* 101: 55-64, 1995.
220. A. Perez. *Clin. Exp. Res.* 19: 714-720, 1995.
221. R. F. Phalen, D. V. Bates. Proceedings of the Colloquim on Partculate Air Pollution and Human Mortality. *Inhalation Toxicol.* 7, 1995.
222. M. A. Philbert. *Neurotoxicology.* 16: 349-362, 1995.
223. A. Poland, A. Kende. *Fed. Proc.* 35: 2404-2408, 1978.
224. T. Prestera, P. T. Alalay. *Proc. Natl. Acad. Sci.* U. S. 92: 8965-8969, 1995.
225. P. Prigent, A. Saoudi, C. P. Parmetier. *J. Clin. Invest.* 96: 1484 1489, 1995.
226. S. Puri, K. K. Kohl. Pharm. *Toxicol.* 77: 136-141, 1995.
227. B. G. Que. *Circulation.* 92: 238 243, 1995.
228. K. Rabe, H. Magnussen, M. Aschner. *J. Neurochem.* 65: 1526-1528, 1995.
229. J. Ralof. *Science News.* 148: 44 47, 1995.
230. V. S. Rana, S. Kumar. *Physiol. Chem. Phys. Med. NMR.* 27: 25-29, 1995
231. I. S. Ratton, G. E. Sibaska, F. P. Wilmar. *Med. Sci. Res.* 23: 469-470, 1995
232. P. K. Ray. *Advances in Immunology and Cancer Therapy.* Springer, Berlin, 1985.
233. W. R. Rea. *Chemical Sensitivity Principles and Mechanisms.* F. L. Lewis, ed. CRC Press, Boca Raton, FL., 1992.
234. K. Redhead. *Pharmacol. Toxicol.* 20: 278-283, 1992.
235. C. A. Reinhardt, C. H. Schein. *Toxicol. in vitro.* 9: 369-374, 1995

236. M. Rempel. In: *Biological Monitoring in Occupational Medicine*. J. LaDan, ed. East NorwaLk, Appleton, CT., 1995, p. 459-466.

237. C. Rice-Evans, B. Halliwell. *Free Radicals, Methodology and Concepts*. Richelieu Press, London, 1988.

238. L. Rising, D. Vitarella, M. Aschner. *J. Neurochem.* 65: 1562-1568, 1995.

239. R. A. Roberts, A. R. Soames. *Carcinogenesis.* 16: 1639, 1995.

240. I. R. Rowland. In: *Role of Gut Bacteria in Human Toxicology and Pharmacology*. M. J. Hill, ed. Taylor Francis, London, 197-211, 1995.

241. J. R. Sabine. *CRC Critical Rev. Toxicol.* 67: 190-198, 1981.

242. S. H. Saff. *Pharmacol. Therap.* 67: 247-251, 1995.

243. A. Saija, M. Scalese, M. Lanza. *Free Rad. Biol. Med.* 19: 481-486, 1995.

244. M. Saito, S. Ikegami, T. Aizawa. *J. Nutr. Vitamin.* 29: 467-451, 1983.

245. H. Saito, S. Iwani, T. Shigeaka. *Toxicol. Environ. Health.* 41: 194-205, 1995.

246. M. Sakar, K. Yarnagani. *Pharmacol. Toxicol.* 77: 36-40, 1995.

247. M. Samanek, A. Zapetel. *Progress in Respiration Research.* vol. 22, Rep. 62, S. Krager, Basel. 1987.

248. T. Sandestroem. *Europ. Respir. J.* 8: 976-995, 1995.

249. U. I. Sanzgiri, H. I. Kim, C. E. Dallas. *Toxicol. Appl. Pharmacol.* 134: 148-154, 1995.

250. T. A. Sarafian, D. E. Bredesen. *Free Rad. Res.* 21: 1-8, 1994.

251. A. H. Sarker, S. Watanabe, S. Seki. *Mutat. Res.* 337: 85-95, 1995.

252. E. N. Schacter, T. J. Witek, G. J. Berth. *Arch. Environ. Health.* 39: 34-42, 1984.

253. P. Scholl, S. Musser. *Pharmacogenetics.* 5: S171-S176, 1995.

254. I. Schraufstatter, S. T. Revak, *Fed. Proc.* 43: 2807-2809, 1987.

255. G. Scott. *Atmospheric Oxidation and Antioxidants*, vol. 1. Elsever, Amsterdam, 1993.

256. A. Seaton, D. J. Gooden, K. Brown. *Thorax.* 49: 171-174, 1994.

257. M. J. K. Selgrade. *Inhalation Toxicol.* 7: 891-900, 1995.

258. I. V. Semak, A. T. Pikulev. *Drug Metab. Drug Interact.* 11: 244-267, 1994.

259. D. S. Sharp, E. A. Hassan. *Arch. Environ. Contam. Toxicol.* 29: 424-428, 1995.

260. P. Shubik. *Cancer Lett.* 93: 3-7, 1995.

261. G. Siest, M. Galteau, F. Schefle. In: *Drug Measurement and Drug Effects*. G. Scest, ed. S. Krager, Basel, Chapter 7, 1987.

262. A. Singh. *Can. J. Physiol. Pharmcol.* 60: 1330-1335, 1982.

263. V. P. Singh. *Prog. Ind. Microbiol.* 32: 51-63, 1995.

264. T. F. Slater. *Biochemical Mechanisms of Liver Injury*. Acad. Press, NY, 1985.

265. T. F. Slater, D. McBrien. *Free Radicals, Lipid Peroxidation and Cancer*. Acad. Press, NY, 128-137, 1982.

266. M. C. Smith, H. Thorpe. *Biochem. Pharmacol.* 16: 763-768, 1986.

267. V. Soseth, J. Kongerud, D. Haar. *Lancet.* 345: 217-220, 1995.

268. G. J. Spies, R H. Rhyne, R. A. Evans, K E. Wetzal. *J. Occup. Env. Med.* 37: 1102-1107, 1995.

269. M. B. Spron, A. B. Roberts. *Cancer Res.* 43: 3034-3037, 1953.

270. N. H. Stacey, C. D. Klaasen. *J. Toxicol. Environ. Health.* 7: 136-139, 1982.

271. H. M. Steinman. In: *Oxygen Radicals in Biology and Medicine*. M. G. Simic, K. A. Taylor, J. F. Ward, eds. Plenum Press, NY, 641-646, 1987.

272. H. F. Stich. *Carcinogens and Mutagens in the Environment*. CRC Press, Boca Raton, FL. 1982.

273. R. Stocker. *Free Rad. Res. Com.* 9: 101-112, 1990.

274. R. Stocker, Y. Yamamoto, B. N. Ames. *Science.* 235: 1043-1046, 1984.

275. S. J. Stohs, D. Bagchi. *Free Rad. Biol. Med.* 18: 321-337, 1995.

276. A. Szet Gyorgyi. *Electronic Biology of Cancer*. M. Dekker, NY, 1976.

277. A. E. Taylor, D. Martin, J. C. Packer. *Surgery.* 94: 433-438, 1983.

278. J. S. Tepper, J. R. Lehmann. *Am. J. Resp. Crit. Care Med.* 149: A839-A842, 1994.

279. J. M. Thayer, P. E. Murkes. *Teratology.* 51: 418-429, 1995.

280. J. Thillet, A. M. Michelson. *Free Rad. Res. Com.* 1: 89-100, 1985.

281. Threshold Limit Values for Chemical Substances and Pharmacological Agents and Biological Exposure. *Am. Conf. Gov. Ind. Hygiene.* Cincinnatti, OH, 1993.

282. R. Thurman, B. U. Bradford, K. T. Knecht. In: *Alcohol Metabolism and its Toxicology*, Alcohol Drug Abuse. R. R. Watson, ed. CRC Press, Boca Raton, FL. 145-161, 1995.

283. D. R. Tisot, M. R. Deray. *Environ. Res.* 44: 71-77, 1987.

284. M. Tornquist, H. H. Landin. *J. Occup. Environ. Med.* 37: 1077-1084, 1995.

285. Y. Uejima, J. Fukuki. *Exp. Lung Res.* 21: 631-642, 1995.

286. T. Velpar, J. Schneider, A. Morin. *Environ. Toxicol. Chem.* 14: 1602-1613, 1995.

287. V. Voicu, R. Olinescu. *Enzymatic Mechanisms in Pharmacodynamics.* Abacus Press, London. 1983.

288. A. A. Walsh, L. A. Graftenreid, R. N. Rice. *Carcinogenesis.* 16: 2187-2189, 1995.

289. J. H. Wang, J. Duddle, J. L. Devalia. *Int. Arch Allergy Immunol.* 107: 103-105, 1995.

290. L. W. Wattenberg. *J. Environ. Phatol. Toxicol.* 3: 373-378, 1981.

291. S. J. Weiss. *Science.* 227: 317-319, 1985.

292. M. Werner, M. J. Costa. *Clin. Chim. Acta.* 237: 107-154, 1995.

293. H. E. Weygand, J. Wu. *Chem. Res. Toxicol.* 8: 955-958, 1995.

294. A. G. Wiese, R. E. Pacifici, K. Adavies. *Archiv. Biochem. Biophys.* 318: 231-240, 1995.

295. A. S. Wilson, M. T. Tringle, M. D. Kelly. *Hum. Exp. Toxicol.* 14: 507-515, 1995.

296. D. A. Wink. *Science.* 254: 1001-1003, 1995.

297. C. C. Winterbourn. In: *Oxygen Radicals Systemic Events and Disease Processes.* D. K. Das, W. B. Easumen, eds. S. Krager, Basel, 1990.

298. A. S. Wright. In: *Developments in Science and Practice of Toxicology.* A. W. Hayes, R. C. Schnell, eds. Elsever, London, 1983, p. 311-318.

299. J. L. Zweier. In: *Oxy-Radicals in Molecular Biology and Pathology.* P. A. Ceruffi, J. M. McCord, eds. A. R. Liss Publ., NY, 1989. p. 268-275.

300. K. Yagi. Active Oxygen, *Lipid Peroxidation and Antioxidants.* CRC Press, Boca Raton, FL., 1993.

301. C. S. Yang, P. J. Tsai, L. L. Liu, J. S. Kuo. *Free Rad. Biol. Med.* 19: 453-459, 1995.

302. S. H. Yuspa. *Diet, Nutrition and Cancer.* A. R. Liss, NY, 1984.